Algebra Two

David Eastwood

PARTRIDGE

Copyright © 2019 by Brain Based Education.
Math Without Calculators
Algebra Two
Medina, Ohio USA

ISBN: Softcover   978-1-5437-0544-7

All rights reserved. No part of this book may be used or reproduced by any means, graphic, electronic, or mechanical, including photocopying, recording, taping or by any information storage retrieval system without the written permission of the author except in the case of brief quotations embodied in critical articles and reviews.

Summary: "This book Is for students who want to learn math, especially those who want to learn it without calculators. It is filled with suggestions and problems to review it."

Because of the dynamic nature of the Internet, any web addresses or links contained in this book may have changed since publication and may no longer be valid. The views expressed in this work are solely those of the author and do not necessarily reflect the views of the publisher, and the publisher hereby disclaims any responsibility for them.

Print information available on the last page.

To order additional copies of this book, contact
Partridge India
000 800 10062 62
orders.india@partridgepublishing.com

www.partridgepublishing.com/india

Algebra 2

9. Polynomials, Asymptotes, and Complex Numbers

1. Basic Polynomials and Graphs.......1
- 1-1 Polynomials/Signs — 1
- 1-2 Graph Even/Odd Exponents — 3
- 1-3 2 Term Equations — 5
- 1-4 Graph Different Exponents — 7
- 1-5 Review Problems — 9

2. How Binomials Work.....................11
- 2-1 Ma/Me Rules — 11
- 2-2 DS Rule/Simplification — 13
- 2-3 Power Rule — 15
- 2-4 Review Problems — 17

3. How Binomails Work Pt 2...............19
- 3-1 Complete/Square — 19
- 3-2 Quadratic Formula — 21
- 3-3 Add/Subtract Cubes — 23
- 3-4 Review Problems — 25

4. Divide Polynomials..........................27
- 4-1 Divide a Box — 27
- 4-2 Missing Term — 29
- 4-3 Synthetic Division — 31
- 4-4 Non Standard Equations — 33
- 4-5 Review Problems — 35

5. Different Theorems........................37
- 5-1 Fundamental Thereom — 37
- 5-2 Remainder/Factor Ther — 39
- 5-3 Graph Binomials — 41
- 5-4 Review Problems — 43

6. Begin Asymptotes...........................45
- 6-1 Asymptotes — 45
- 6-2 Binomial Denominator — 47
- 6-3 Numerator Binomial — 49
- 6-4 Review Problems — 51

7. Asymptotes Pt 2...............................53
- 7-1 Discontinuity — 53
- 7-2 Binomial Discontinuity — 55
- 7-3 Asymptotes Word Problems — 57
- 7-5 Review Problems — 59

8. Rational Equations..........................61
- 8-1 Direct and Indirect Variation — 61
- 8-2 Joint Variation — 63
- 8-3 Multiply Fractions Exponents — 65
- 8-4 Divide/Complex Fractions — 67
- 8-5 Review Problems — 69

9. Binomial Fractions Pt 2...................71
- 9-1 Add Binomial Denominators — 71
- 9-2 Complex Fractions — 73
- 9-3 How to Change an Equation — 75
- 9-3 Review Problems — 77

10. Polynomials and Roots.................79
- 10-1 Match Exponents — 79
- 10-2 Multiply Roots — 81
- 10-3 DS/Simplification — 83
- 10-4 Fractions Exponents — 85
- 10-5 Review Problems — 87

11. Roots Pt 2..............................89
    11-1 Change Base 2 Ways   89
    11-2 Square Roots in Equations 91
    11-3 Review Problems   93

12. Complex Numbers.............................95
    12-1 Complex Numbers   95
    12-2 Add/Subtract Binomials   97
    12-3 Multiply Complex Numbers 99
    12-4 Review Problems   101

13. Complex Numbers Pt 2..................103
    13-1 Larger Exponents   103
    13-2 Simplify Fractions   105
    13-3 Equations   107
    13-4 Story Problems   109
    13-5 Review Problems   111

14. Polynomial Functions....................113
    14-1 Polynomials   113
    14-2 Add Functions   115
    14-3 Piecewise Functions   117
    14-4 Compose Functions   119
    14-5 Review Problems   121

15. Inverse Functions.........................123
    15-1 Inverse Functions   123
    15-2 Point and Line Inverses   125
    15-3 Inverse Quadrilaterals   127
    15-4 Review Problems   129

Algebra 2
Matrices, Conics, Probability, Sequences, and Series

1. Matrices and Geometry..............131

1-1	Begin Matrices	131
1-2	With Geometry	133
1-3	Geometry with Shapes	135
1-4	Add Matrices	137
1-5	When to Multiply	139
1-6	Multiply 2 x 1, 2 x 2	141
1-7	Review Problems	143

2. Matrices and Equations...............145

2-1	Multiply Equations	145
2-2	Cramer's Law	147
2-4	Cramers Rule	149
2-5	3 Variable Systems	151
2-6	Review Problems	153

3. Circle Equations and Graphs......155

3-1	Circles Review	155
3-2	Move a Circle with H, K	155
3-3	Is a Point on a Circle	157
3-4	Distance Formula	159
3-5	Review Problems	161

4. Circles Pt 2....................163

4-1	Find X, Y Intercepts	163
4-2	Complete the Square	165
4-3	Midpoint Problems	167
4-4	Make Equations from Graph	169
4-5	Review Problems	171

5. Ellipse Equations and Graphs.....173

5-1	Begin Ellipses	173
5-2	Draw a Foci Equation	175
5-3	B and C Equations	177
5-4	Ellipse Equations	179
5-5	Complete the Square	181
5-6	Area/Perimeter of an Ellipse	183
5-7	Review Problems	185

6. Hyperbola Equations / Graphs.....187

6-1	Begin Hyoerbole	187
6-2	B Formula	189
6-3	Asymptotes Box and C	191
6-4	Hyperbole Moved from 0, 0	193
6-5	Review Problems	195

7. Parabola Systems and Graphs....197

7-1	Parabolas	197
7-2	Quick Formula	199
7-3	Substitute Systems	201
7-4	Elimination Systems	203
7-5	Review Problems	205

8. Probability and Events.................207

8-1	Fractions/Percents	207
8-2	2 Events	209
8-3	Permutations	211
8-4	Combinations	213
8-5	Review Problems	215

9. Two Step Probabilities................217

    9-1 Overlap Probabilites   217
    9-2 Multiple Combinations   219
    9-3 Multiply Permutations   221
    9-4 Review Problems   223

10. Standard Dev/Box Whiskers....225

    10-1 Standard Deviations   225
    10-2 With Calculator   227
    10-3 Box Whiskers   229
    10-4 Practice Box Whiskers   231
    10-5 Review Problems   233

11. Binomial Theorem and Others..235

    11-1 Margin of Error   235
    11-2 Central Tendency   237
    11-3 Normal Distribution   239
    11-4 Binomial Thereom   241
    11-5 Scatter Plots   243
    11-6 Linear Programming   245
    11-7 Review Problems   247

12. Arithmetic Sequences/Series....249

    12-1 Begin Sequences   249
    12-2 Sequence Problems   251
    12-3 Series Formula   253
    12-4 Notation/Story Problems   255
    12-5 Review Problems   257

13. Geometric/Infinite Seq/Series....259

    13-1 Geometric Sequence   259
    13-2 Negatives/Fractions   261
    13-3 Geometric Series   263
    13-4 Story Problem   265
    13-5 Review Problems   267

14. Infinite Series...........................269

    14-1 Infinite Series Formula   269
    14-2 Use Sigma Sign   271
    14-3 Fibonacci/Fractal   273
    14-4 Review Problems   275

Algebra 2
Logarithms and Trigonometry

1. Basic Logarithm Exponents....277

    1-1 Begin Exponents   277
    1-2 Logarithm Formula   279
    1-3 Exponential Formula   281
    1-4 2 to 4 and 3 to 6 digitis   283
    1-5 7, 8, 9, and 5   285
    1-6 Backwards   287
    1-7 Review Problems   289

2. Add, Subtract, Fraction Logs..291

    2-1 Add Logarithms   291
    2-2 Largers Logarithms   293
    2-3 Subtract Logarithms   295
    2-4 Fraction Logs   297
    2-5 Fractions less than 1   299
    2-6 Log of Decimals   301
    2-7 Review Problems   303

3. Two Digit/Power Logarithms...305

    3-1 2 Digit Logarithms   305
    3-2 Scientific Notation   307
    3-3 Power Logarithms   309
    3-4 Radical Logarithms   311
    3-5 Review Problems   313

4. Other Bases/Exponential Eq...315

    4-1 Other Bases   315
    4-2 Change to Base 10 (others)   317
    4-3 Exponential Equations   319
    4-4 Same Base Equations   321
    4-5 Story Problems   323
    4-6 Review Problems Pt 2   325

5. Logarithm Equations.............327

    5-1 Log Equations
    5-2 Antilogarithms   327
    5-3 Take Log Both Sides   329
    5-4 MA Rule Equations   331
    5-5 DS Rule Equations   333
    5-6 Review Problems   335

6. Natural Numbers....................337

    6-1 Natural Numbers   337
    6-2 Continuous Logarithms   339
    6-3 Natural Logarithms   341
    6-4 Review Problems   343

7. Logarithm Story Problems...345

    7-1 Earthquakes   345
    7-2 Compare Earthquakes   347
    7-3 PH Levels   349
    7-4 Decibels with Logarithms 351
    7-5 Story Problems   353
    7-6 Review Problems Pt 2   355

8. Angles and Radians..............357

    8-1 Angle Basis   357
    8-2 Radians Under Pi   359
    8-3 Degree to Radians   361
    8-4 Radians to Degraees   363
    8-5 Revolutions   365
    8-6 Story Problems   367
    8-7 Review Problems   369

9. Sin Ratios and Triangles....................371

    9-1 Triangles Labeled   371
    9-2 Sin Angles to 36 degrees   373
    9-3 Sin Angles over 36   375
    9-4 Square Root Answers   377
    9-5 Story Problems   379
    9-6 Review Problems   381

10. Cosine and Tangent Ratios..............383

    10-1 Cosine Angles   383
    10-2 Cosine Angles from 0   385
    10-3 Tangents   387
    10-4 Tangent Story Problems   389
    10-5 Story Problems   391
    10-6 Review Problems   393

11. Secants, Cosecants, Cotangents....395

    11-1 Cosecants   395
    11-2 Secants   397
    11-3 Cotangents   399
    11-4 Story Problems   401
    11-5 Review Problems   403

12. Sin and Cosine Laws........................405

    12-1 Area of a Triangle   405
    12-2 Sine Laws   407
    12-3 Triangle Signs   409
    12-4 Law of Cosines   411
    12-5 Centripetal Acceleration   413
    12-5 Story Problems   415
    12-5 Review Problems   417

13. Circular Functions...........................419

    13-1 Circular Functions   419
    13-2 Tangential/Change the Graph   421
    13-3 Change the Graph   423
    13-4 Change the Graph Pt 2   425
    13-4 Review Problems   427

Ch 4 Ls 1: 2 Steps to Carry Addition   33

Why we're different!!!

_____ Front ____ / 8 Back ____ / 27 Rev ____ / 20 T / 53 _____
         Name   Checker

#1 1. When do you carry in math? _____
          26
2. What does 6 + 6 carry? + 6 _____

3. Signs here when they're done.

3. What's the 2nd step to carry? _____

4. Why don't you have to carry twice with 100s? _____

#2 1. You know 3 + 9 is 12.
     What's the 1st step to carry?

    13
    + 9
    ———
     2

1. Student makes sure these are filled out in class.

    → ¹1 3
Carry a ____.
    + 9
    ———
     2

___ ___
tens ones

2. What's the teen fact?
    15
    + 9

2. Checker quizzes the student on the front page.

2. What's the teen fact?
    15
    + 9

5 + 9 = ____   Carry a 10.
                 How many 10s in all?

___ ___

3. Find the teen fact.
    34
    + 8

4 + 8 = ____   Carry a 10.
                 How many 10s in all?

___ ___

34.

#3 We did teens. Finish them.   Calculator?
                                                              yes   no

1. 13   15   18   45   56
    + 8   + 8   + 6   + 5   + 5
    ―――   ―――   ―――   ―――   ―――
     1   3   4   0   1   ┌─────────────────────┐
                                               │ 4. Student does   │
2. 26   54   45   37   74   │   these problems. │
    + 6   + 8   + 5   + 5   + 7   └─────────────────────┘

3. 15   15   16   32   54
    +15   +25   +18   +19   +19
    ―――   ―――   ―――   ―――   ―――

              #4 Is it correct? Yes or correct it.   Calculator?
                                                              yes   no
1. 43   54   65   39
    + 7 yes   + 7 yes   + 5 yes   + 3 yes
    ―――   ―――   ―――   ―――
    51 ___   61 ___   60 ___   43 ___

2. 29   46   54   22
    +19 yes   +37 yes   +28 yes   +29 yes
    ―――   ―――   ―――   ―――
    38 ___   84 ___   72 ___   50 ___

3. 57   45   19   36
    +35 yes   +16 yes   +25 yes   +34 yes
    ―――   ―――   ―――   ―――
    82 ___   61 ___   44 ___   70 ___

Review 1. When do you carry in math? _____   Calculator?
                                                                  yes   no
        2. Name 2 steps to carry. _____

        3. Why don't you have to carry twice with 100s? _____

4. 19   18   16   32   64   ┌─────────────────────┐
    + 3   + 7   + 5   + 9   + 8   │ 5. Student is   │
                                               │   quizzed on   │
5. 28   45   55   76   74   │   these Qs.   │
    + 4   + 7   + 5   + 6   + 8   └─────────────────────┘

                                              (Teacher option)

                      ┌──────────────────┐
                      │ It works!!!   │
                      └──────────────────┘

Rules Sheet

As you go through the lessons you will see some rules that you are unfamiliar with, so I developed this paper to help you understand them.

1. Ma Rule: The intials are from the rule itself, "Multiply same bases, Add the exponents." You probably know the Product of Powers Property.
   Example: $2^2 \times 2^4 = 2^6$

2. Me Rule: These initials are also from the rule, "Multiply the numbers, Exponents stay the same." (E shows that exponents stay the same.)
   Example: $3^4 \times 5^4 = 15^4$

3. DS Rule: Again, the initials show the rule. "Divide same bases, Subtract the Exponents."
   Example: 4^5 divided by $4^2 = 4^3$

4. Simpliify: No need for initials here, just simplify.
   Example: 20^3 divided by $4^3 = 5^3$

5. Subtraction: 2. Count Up Subtraction: Example: 23 - 9 Count Up 9 to 10 and subtract, it's 13. Add the 1 from the bottom. It's 14.

These are the name changes up until now. Any others?

Ch 1 Ls 1 Solve with even and odd exponents. 1

_____ #1 #2 ____ / 11 #3 ____ / 12 R ___ / 5 T ___ / 41 _____
    Name   Checker

#1 1. What makes a quadratic equation? _____
       2. What sign does a negative quadratic make? _____
       3. What makes a cubic equation? _____
       4. What sign does a negative cube make? _____
       5. If an equation has the term x^2y^3, how does it find the degree? _____

#2 1. Solve for x is 2 and - 2. $y = x^3 + 1$

                                        2 is _____   - 2 is _____

       2. Solve for x is $\frac{1}{2}$ and - $\frac{1}{2}$. $y = x^3 + 1$

                                        1/2 is _____   - 1/2 is _____

       3. Solve for x is 0.2 and - 0.2. $y = x^3 + 1$

                                        0.2 is _____   - 0.2 is _____

       4. Solve for x is 2 and - 2. $y = -x^4 + 1$

                                        2 is _____   - 2 is _____

       5. What is the degree of this equation? $y = a^3 b^4 + 1$

                                            _____

       6. What is the degree? $y = x^4 z^4 + 1$

                                            _____

2.

#3 What is the degree of each equation?   Calculator?
yes   no

1. $y = a^3 b^4 + 1$ _____ $y = x^4 y^4 + 2$ _____
2. $y = a^4 b^2 + 6$ _____ $y = x^7 y^2 + 3$ _____
3. $y = a^5 b^4 + 4$ _____ $y = x^5 y^3 + 5$ _____
4. $y = a^9 b^3 + 1$ _____ $y = x^6 y^4 + 7$ _____

#4 Determine what the value of each equation is   Calculator?
yes   no

1. Solve for x is 2 and -2. $y = x^3 + 1$ (2, ____) (-2, ____)
   $y = -x^4 + 1$ (2, ____) (-2, ____)

2. Solve for x is $\frac{1}{2}$ and $-\frac{1}{2}$. $y = 2x^3 + 4$ (0.5, ____) (-0.5, ____)
   $y = 2x^4 + 4$ (0.5, ____) (-0.5, ____)

3. Solve for x is 2 and -2. $y = x^5 - 2$ (2, ____) (-2, ____)
   $y = x^6 - 2$ (2, ____) (-2, ____)

4. Solve for x is 3 and -3. $y = x^3 + 3$ (3, ____) (-3, ____)
   $y = x^4 + 3$ (3, ____) (-3, ____)

Review   1. What makes a quadratic equation? _____   Calculator?
yes   no
2. What sign does a negative quadratic make? _____
3. What makes a cubic equation? _____
4. What sign does a negative cube make? _____
5. If an equation has the term $x^2 y^3$, how does it find the degree? _____
_____

Ch 1 Ls 2   How even and odd exponents graph.   3

_____ #1 #2 ____/ 7   #3 ____/ 8   R ___/ 5   T____/ 20   _____
           Name   Checker

#1 1. Which term decides the direction for a graph? _____

2. Even exponent graphs make what kind of graph? _____

3. Odd exponents make what kind of graph? _____

4. How does a negative change an odd exponent graph? _____

5. Name 3 steps to predict a graph. _____

_____

#2   1. Find x is 2 and - 2. $y = x^3 + 2$

Graph it.   2, ____   - 2, ____   What's the y intercept, direction, and slope?

Y intercept: _____

Direction: _____

Slope: _____

2. Find x is 2 and - 2. $y = -x^3 - 4$

Graph it.   2 is ____   - 2 is ____   What's the y intercept, direction, and slope?

Y intercept: _____

Direction: _____

Slope: _____

4.

#3 Find the points and graph.

Find 2 points and graph.

1. $y = x^3 + 1$

Y intercept: _____
Direction: _____
Slope: _____

2 ____
-2 ____

$y = 2x^4 - 15$

Y intercept: _____
Direction: _____
Slope: _____

2 ____
-2 ____

Calculator? yes   no

2. $y = -3x^3 + 5$

Y intercept: _____
Direction: _____
Slope: _____

2 ____
-2 ____

$y = -2x^4 - 4$

Y intercept: _____
Direction: _____
Slope: _____

2 ____
-2 ____

Review

1. Which term decides the direction for a graph? _____
2. Even exponent graphs make what kind of graph? _____
3. Odd exponents make what kind of graph? _____
4. How does a negative change an odd exponent graph? _____
5. Name 3 steps to predict a graph. _____
_____

Calculator? yes   no

Ch 1 Ls 3   Same sign equations and graphs.   5

_____ #1 #2 ____/ 9   #3 ____/ 8   R ____/ 5   T ____/ 22 _____
            Name   Checker

#1 1. What does it mean by same sign exponents? _____

2. How do same exponents change a graph? _____

3. What kind of graph does this make? $y = x^3 - 2x$ _____

4. What made it become an S curve? _____

5. How does How Much change a graph? _____

#2 1. Compare terms, then graph.
Solve for x is 2, - 2, 3, and - 3. $y = -x^3 + 4x$

2, ____   - 2, ____   3, ____   - 3, ____

Graph it. Is it an S curve or not?
Relative maximum or minimums?

S Curve   Yes   No

Relative maximum or minimums?

Yes   No

2. Compare terms, then graph.
Solve for x is 2, - 2, 3, and - 3. $y = -x^3 + 3x$

2, ____   - 2, ____   3, ____   - 3, ____

Graph it. Is it an S curve or not?
Relative maximum or minimums?

S Curve   Yes   No

Relative maximum or minimums?

Yes   No

6.

#3 Graph each of the equations.   Calculator?
                                                                    yes   no

1. $y = 2x^3 - 4x$ $y = -x^4 + 3x$

(1, ___) (2, ___) (-1, ___) (-2, ___)   (1, ___) (2, ___) (-1, ___) (-2, ___)

Solve x is 1, 2 and -1 -2.
Same or different signs?

Graph them. Find relative
maximum or minimums
and if there's an S curve.

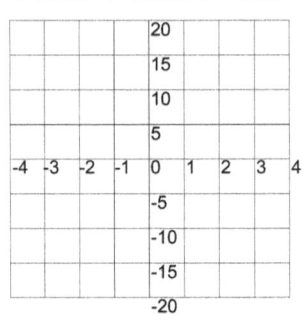   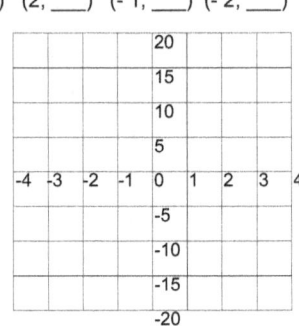

2. $y = -x^3 + 5x$ $y = 2x^4 - 4x$

(1, ___) (2, ___) (-1, ___) (-2, ___)   (1, ___) (2, ___) (-1, ___) (-2, ___)

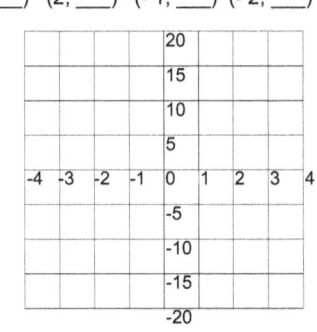

Review 1. What does it mean by same sign exponents? _____

2. How do same exponents change a graph? **3** _____

3. What kind of graph does this make? $y = x - 2x$ _____

4. What made it become an S curve? _____

5. How does How Much change a graph? _____

Ch 1 Ls 4   Same sign equations and graphs.   7

_____#1 #2 ____/ 8 #3 ____/ 8 R ___/ 4 T ___/ 20 _____
            Name   Checker

Example: $y = x^4 + x^3$

#1   1. What are different sign exponents? _____
     2. What happens to the signs for different exponent terms? _____
     3. How is this graph different from cubic graphs? _____
     4. Which term decides the direction of the graph? _____

#2   1. Compare terms, then graph.
        Solve for x is 1, 2 and -1, -2. $y = x^3 - 2x^2$

     Graph it. Is it an S curve or not?
     Relative maximum or minimums?   (1, ___) (2, ___) (-1, ___) (-2, ___)

             S Curve   Yes   No

        Relative maximum or minimums?
                Yes   No

     2. Solve for x is 1, 2 and -1, -2. $y = 2x^4 - 3x^3$

     Graph it. Is it an S curve or not?
     Relative maximum or minimums?   (1, ___) (2, ___) (-1, ___) (-2, ___)

             S Curve   Yes   No

        Relative maximum or minimums?
                Yes   No

8

#3 Solve for x is 1, 2 and -1, -2. Same or different signs?   Calculator? yes   no

1. $y = x^4 - 4x^3$ $y = -x^4 + 5x^3$

(1, ___) (2, ___) (-1, ___) (-2, ___)   (1, ___) (2, ___) (-1, ___) (-2, ___)

Graph them. Find relative maximum or minimums.

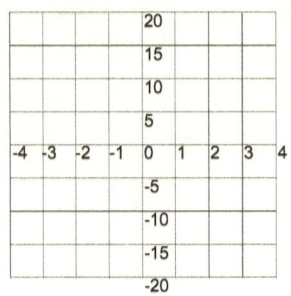
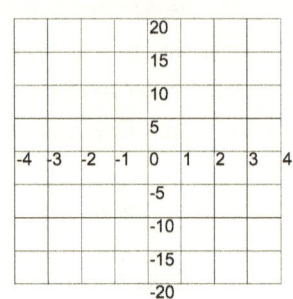

2. $y = -x^3 + 3x^2$ $y = 2x^3 - 7x^2$

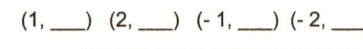

(1, ___) (2, ___) (-1, ___) (-2, ___)   (1, ___) (2, ___) (-1, ___) (-2, ___)

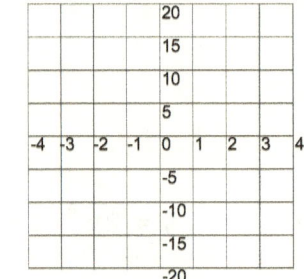
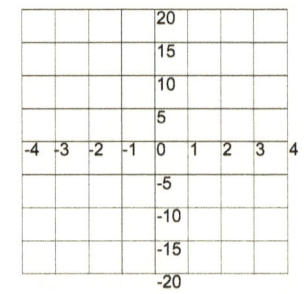

Review 1. What are different sign exponents? _____

_____

2. What happens to the signs for different exponent terms? _____

3. How is this graph different from cubic graphs? _____

4. Which term decides the direction of the graph? _____

Review Problems 9

_____ #1 - #4 _____/ 18   Back _____/ 12   Total _____/ 30
Name

1. Polynomials _____

2. Same Sign Terms _____

3. Different Sign Terms _____

4. S Curve _____

#2 What is the degree of each?

1. $y = a^3 b^4 + 1 =$ _____ $y = x^4 y^4 + 2 =$ _____   Calculator? yes   no

2. $y = a^4 b^2 + 6 =$ _____ $y = x^7 y^2 + 3 =$ _____

3. $y = a^5 b^4 + 4 =$ _____ $y = x^5 y^3 + 5 =$ _____

4. $y = a^9 b^3 + 1 =$ _____ $y = x^6 y^4 + 7 =$ _____

#3 Determine the value.

1. Solve for x is 2 and - 2. $y = x^3 + 1$ (2, ____) (-2, ____)
   $y = -x^4 + 1$ (2, ____) (-2, ____)

2. Solve for x is $\frac{1}{2}$ and $-\frac{1}{2}$. $y = 2x^3 + 1$ (0.5, ____) (-0.5, ____)
   $y = 2x^4 + 1$ (0.5, ____) (-0.5, ____)

#4 Find 2 points and graph.

Find the points and graph.

1. $y = x^3 + 1$

Y intercept: _____
Direction: _____
Slope: _____

2 ____
-2 ____

$y = 2x^4 - 15$

Y intercept: _____
Direction: _____
Slope: _____

2 ____
-2 ____

#5

1. $y = -x^3 + 5x$ $y = 2x^4 - 4x$ Calculator? yes   no

(1, ___) (2, ___) (-1, ___) (-2, ___)   (1, ___) (2, ___) (-1, ___) (-2, ___)

Find relative maximum or minimums.

2. $y = x^4 - 2x^3$ $y = -x^4 + 2x^3$

(1, ___) (2, ___) (-1, ___) (-2, ___)   (1, ___) (2, ___) (-1, ___) (-2, ___)

3. $y = -x^3 + 3x^2$ $y = 2x^3 - 3x^2$

(1, ___) (2, ___) (-1, ___) (-2, ___)   (1, ___) (2, ___) (-1, ___) (-2, ___)

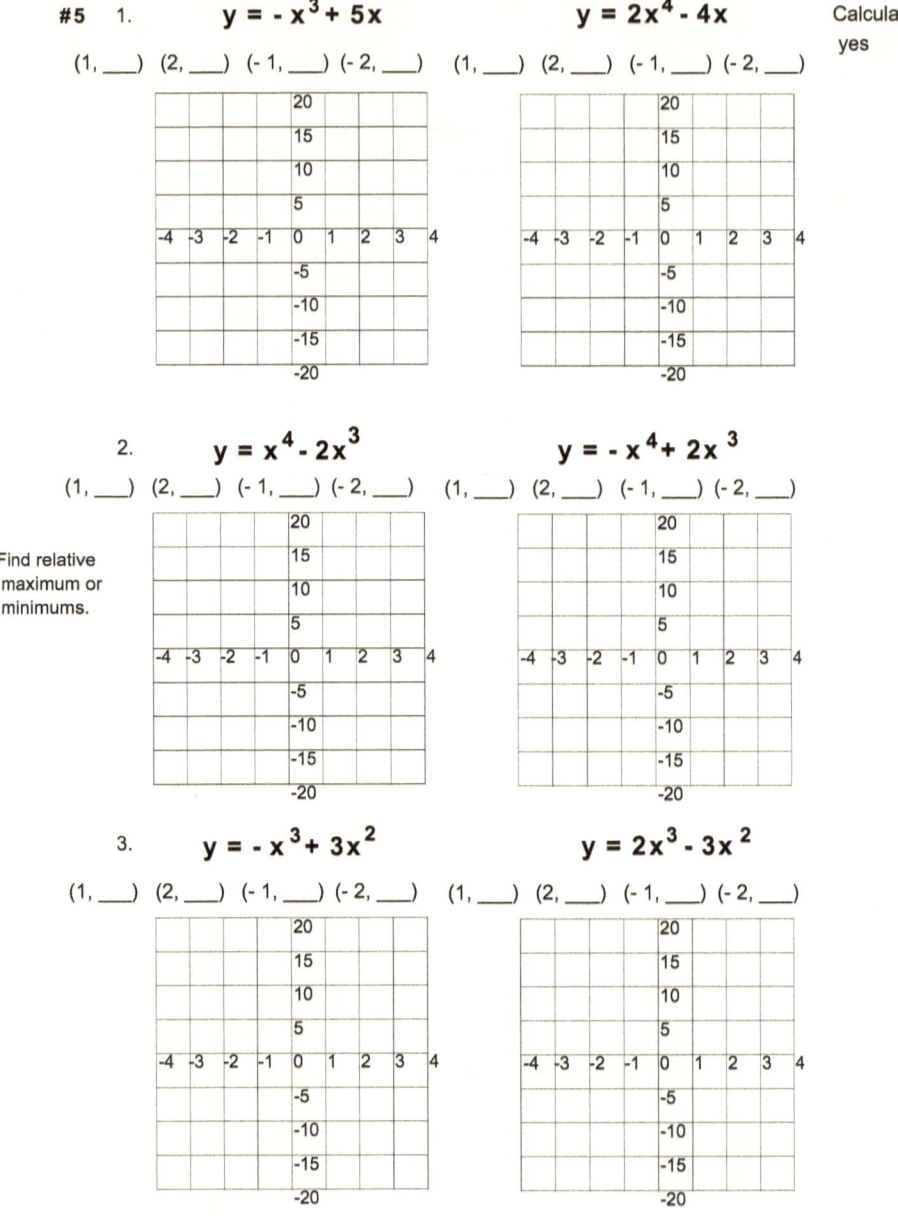

Ch 2 Ls 1   2 ways to multiply with binomials.   11

_____ #1 #2 ____/ 10 #3 ____/ 12 R ____/ 9 T____/ 31 _____
         Name   Checker

#1 1. What is the order to multiply binomials? _____
   2. Name 2 ways numbers on the A term changes how they multiply. _____
   _____
   3. (a - 1)(a - 5) What happens with 2 negatives? _____
   4. (x + 3)(x - 3) What happens with opposite signs? _____

        5. Multiply in 1 step.   (a + 2)(a + 4)

                                                 _____

        6. Multiply the binomials.   (2x + 2)(x - 3)

                                                 _____

        7. Multiply with negatives.   (3b - 2)(2b - 4)

                                                 _____

#2 1. What's the 2nd way to multiply? _____

       2. What's the 1st step? $(a + 2)(a^2 + a + 4)$

          Choose a 2nd step.   ____ ____ ____

                                     ____ ____ ____

                                       _____

       3. What's the 1st step? $(x^2 + 5)(x^2 + 2x + 3)$

          Choose a 2nd step.   ____ ____ ____

                                     ____ ____ ____

                                       _____

#3 Multiply.

1. (x + 2)(3x + 4)   (a - 3)(2a + 4)   (2b - 2)(3b - 4)   Calculator? yes   no
   _____   _____   _____

2. (c + 2)(0.5c - 3)   (x - 1)(4x + 6)   (a - 3)(2a - 4)
   _____   _____   _____

3. (2a - 3)(4a + 1)   (- 3x - 2)(x - 4)   (b - 2)(b - 4)
   _____   _____   _____

4. (c + 4)(c - 4)   (2a + 2)(2a - 3)   (- d - 2)(- d - 4)
   _____   _____   _____

Review

1. What is the order to multiply binomials? _____   Calculator?
2. Name 2 ways numbers on the A term changes how they multiply. _____   yes   no
   _____
3. (a - 1)(a - 5) What happens with 2 negatives? _____
4. (x + 3)(x - 3) What happens with opposite signs? _____
5. What's the 2nd way to multiply? _____

6. $(2a - 3)(a^2 + a + 4)$ $(x + 2)3x^2 + 5x + 2)$

Multiply 1 term at a time.

___ ___ ___   ___ ___ ___
___ ___ ___   ___ ___ ___
_____   _____

7. $(c^2 + 2)(c^2 + 2c + 3)$ $(b^2 + 1)3b^2 + b + 5)$

___ ___ ___   ___ ___ ___
___ ___ ___   ___ ___ ___
_____   _____

Ch 2 Ls 2 How to factor with binomials. 13

_____ #1 #2 ____/ 11 #3 ____/ 6 R ____/ 14 T____/ 31 _____
         Name   Checker

#1 1. $x^2 + 6x + 5$ What's the 1st step to factor a trinomial equation? _____
    2. What's the 2nd step to factor a trinomial? _____
    3. How do you know it works? _____
    4. How does it change if C term is a negative? $x^2 - 2x - 3$ _____
    5. How does the B term show you which set it is? _____

    6. What are the C term factors? $x^2 + 4x + 3$

                                                       (x ___)(x ___)

    7. How does the A term change it? $2x^2 + 5x + 3$

      Which B terms work out?   (2x ___)(x ___)   (2x ___)(x ___)

                                                     ____ + ____ = 5x

    8. Factor the trinomial. $3x^2 + 8x - 3$

      Which B terms work out?   (3x ___)(x ___)   (3x ___)(x ___)

                                                     ____ + ____ = 5x

#2 1. How do you factor seperate binomials? _____

    2. Factor each group seperately. $2x^3 + 4x^2 + 9x + 3$

                                                     _____
                                                     _____

    3. Factor each group seperately. $5x^3 + 6x^2 - 2x - 1$

                                                     _____
                                                     _____

#3 Factor each group seperately.   Calculator? yes no

1. $4x^3 + 6x^2 + 6x + 3$ $2x^3 + 6x^2 + 2x + 4$

_____   _____

_____   _____

2. $3x^3 + 9x^2 + 4x - 4$ $-3x^3 - 6x^2 + 8x + 6$

_____   _____

_____   _____

3. $3x^3 + 4x^2 - 6x - 4$ $8x^3 + 2x^2 + 4x + 2$

_____   _____

_____   _____

Review 1. $x^2 + 6x + 5$ What's the 1st step to factor a trinomial equation? _____   Calculator? yes no
2. What's the 2nd step to factor a trinomial? _____
3. How do you know it works? _____
4. How does it change if C term is a negative? $x^2 - 2x - 3$ _____
5. How does the B term show you which set it is? _____
6. How do you factor seperate binomials? _____

How do you factor these?

7. $2x^2 + 7x + 3$ $3x^2 + 7x + 4$

_____   _____

8. $4x^2 + 12x + 5$ $2x^2 + 17x + 8$

_____   _____

9. $3x^2 - 10x - 8$ $4x^2 - 4x - 3$

_____   _____

10. $5x^2 + 2x - 3$ $6x^2 - 3x - 3$

_____   _____

Ch 2 Ls 3 Splitting the Middle Term. 15

_____ #1 #2 ____ / 9 #3 ____ / 6 R ____ / 12 T ____ / 27 _____
Name   Checker

$$6x^2 - 13x + 6$$

#1 1. How do you split a middle number? _____
    2. What makes the 1st part of the answer? _____
    3. What's the 2nd step? _____
    4. Is the middle term added or multiplied? _____

#2 1. How do you split 10x? $\quad 3x^2 + 10x + 8$

$\qquad\qquad\qquad\qquad 3x^2 + 6x + 4x + 8 \qquad$ Split the left side.

$\qquad\qquad\qquad\qquad$ _____ $\qquad$ Split the right side.

$\qquad\qquad\qquad\qquad$ _____ $\qquad$ What are the possible answers?

$\qquad\qquad\qquad\qquad$ _____ $\qquad$ What is the answer?

$\qquad\qquad\qquad\qquad$ _____

    2. How do you split 11x? $\quad 12x^2 + 11x - 15$

$\qquad\qquad\qquad\qquad 12x^2 + 20x - 9x - 15 \qquad$ Split both sides.

$\qquad\qquad\qquad$ _____ _____ $\qquad$ What are the possible answers?

$\qquad\qquad\qquad\qquad$ _____ $\qquad$ What is the answer?

$\qquad\qquad\qquad\qquad$ _____

    3. How do you split 12x? $\quad 6x^2 + 12x + 10$

$\qquad\qquad\qquad\qquad 6x^2 + 2x + 10x - 10 \qquad$ Split both sides.

$\qquad\qquad\qquad$ _____ _____ $\qquad$ What are the possible answers?

$\qquad\qquad\qquad\qquad$ _____ $\qquad$ What is the answer?

$\qquad\qquad\qquad\qquad$ _____

#3 **What's the middle number? Factor it.**

1. $3x^2 + 18x + 4$ $2x^2 + 36x + 6$ Calculator? yes no

2. $4x^2 + 32x - 4$ $3x^2 + 18x + 6$

3. $5x^2 + 30x + 2$ $4x^2 + 72x + 3$

Review
1. How do you split a middle number? _____   Calculator? yes no
2. What makes the 1st part of the answer? _____
3. What's the 2nd step? _____
4. Is the middle term added or multiplied? _____

5. $x^2 - 11x - 40$ $2x^2 + 10x - 16$

6. $4x^2 + 14x + 24$ $x^2 + 12x - 36$

7. $x^2 + 12x - 22$ $2x^2 + 8x - 12$

8. $3x^2 - 12x + 8$ $2x^2 - 17x - 30$

Review Problems   17

_____ #1 - #3 _____ / 17   Back _____ / 14   Total _____ / 31
Name

1. Subtracted Cubes _____

2. Added Cubes _____

#2 1. (x + 2)(3x - 4)   (a - 3)(2a + 4)   (2b - 2)(3b - 4)

Multiply   _____   _____   _____

2. (2c + 2)(1.5c - 3)   (x - 1)(4x + 6)   (a - 3)(2a - 4)

_____   _____   _____

3. (2a - 3)(4a + 1)   (-4x - 2)(x - 4)   (b - 2)(b - 4)

_____   _____   _____

#3 1. $(2a - 3)(a^2 + a + 4)$ $(x + 2)(3x^2 + 5x + 2)$

Multiply the
2nd way.

2. $(4b - 5)(b^2 + 2b + 1)$ $(5x + 4)(3x^2 + 9x + 8)$

3. $(3c + 4)(2a^2 + 6a + 8)$ $(x + 6)(3x^2 + 7x - 3)$

#4 Factor each group seperately.   Calculator? yes no

1. $2x^3 + 4x^2 + 9x + 3$ $2x^3 + 6x^2 + 2x + 4$

   _____   _____
   _____   _____

2. $3x^3 + 9x^2 + 4x - 4$ $-3x^3 - 6x^2 + 8x + 6$

   _____   _____
   _____   _____

#5 What's the middle number? Factor it.   Calculator? yes no

1. $3x^2 + 18x + 4$ $2x^2 + 36x + 6$

   _____   _____
   _____   _____

2. $4x^2 + 32x - 4$ $3x^2 + 18x + 6$

   _____   _____
   _____   _____

3. $5x^2 + 30x + 2$ $4x^2 + 72x + 3$

   _____   _____
   _____   _____

4. $5x^2 + 40x + 8$ $3x^2 + 48x + 8$

   _____   _____
   _____   _____

5. $2x^2 + 36x - 6$ $4x^2 + 24x + 8$

   _____   _____
   _____   _____

Ch 3 Ls 1 Complete the square review. 19

_____ #1 #2 ____ / 7 #3 ____ / 6 R ____ / 5 T ____ / 18 _____
      Name   Checker

#1 1. What does Complete the Square do? _____
        2. What does CB22 mean? _____
        3. What does BS tell you about the new C? _____
        4. How does it always factor? _____
        5. What does SS2 in FSS2 show? _____

#2 1.   What's the 1st step? $x^2 + 6x = 0$

              2 steps to find a new C.   _____

              What happens to it?   _____

              What does F in FSS2 do?   _____

              What is the positive answer?   _____

              Find the negative answer.   _____

                                              _____

        2.   What happens to the + 1? $x^2 + 4x + 1 = 0$

              2 steps to find a new C.   _____

              What happens to it?   _____

              What does F in FSS2 do?   _____

              What is the positive answer?   _____

              Find the negative answer.   _____

                                              _____

#3 Use these equations to create another equation.   Calculator?
                                                              yes   no

1. $x^2 + 2x + 3 = 0$ $x^2 + 6x + 2 = 0$

    _____   _____

    Divide: ___ Square: ___   Divide: ___ Square: ___

New Eq _____   _____

Factor _____   _____

        _____   _____

2. $x^2 + 4x + 7 = 0$ $x^2 + 4x + 5 = 0$

    _____   _____

    Divide: ___ Square: ___   Divide: ___ Square: ___

New Eq _____   _____

Factor _____   _____

        _____   _____

3. $x^2 + 10x + 4 = 0$ $x^2 + 2x + 6 = 0$

    _____   _____

    Divide: ___ Square: ___   Divide: ___ Square: ___

New Eq _____   _____

Factor _____   _____

        _____   _____

Review 1. What does Complete the Square do? _____
        2. What does CB22 mean? _____
        3. What does BS tell you about the new C? _____
        4. How does it always factor? _____
        5. What does SS2 in FSS2 show? _____

Ch 3 Ls 2 Solve the Quadratic Formula. 21

_____ #1 #2 ____/ 7 #3 ____/ 3 R ____/ 6 T____/ 16 _____
          Name   Checker

#1 1. What does the quadratic formula do? _____
   2. What formula starts the quadratic formula? _____
   3. What is the formula for the Discrirminant? _____
   4. Name 4 steps to solve a quadratic formula. _____

_____

#2 1. What's the positive answer? $\dfrac{-2 \pm \sqrt{1}}{2(1)}$

   Find the negative answer.   _____

   2. What's the positive answer? $\dfrac{-6 \pm \sqrt{100}}{2(2)}$

   Find the negative answer.   _____

   3. What's the discriminant? $y = 2x^2 + 3x + 5$

   Solve the discrinant.   ____ - 4( ) ( )

   What's the positive answer?   _____

   Find the negative answer. $\dfrac{- \pm \sqrt{}}{2()} = $ ____

   $\dfrac{- \pm \sqrt{}}{2()} = $ ____

#3 Solve these with quadratic formulas.   Calculator? yes no

1. $y = 2x^2 + 3x + 5$

   ___ - 4( )( )

   _____

   $\dfrac{-\ \pm\sqrt{}}{2(\ )} =$ ___

   $\dfrac{-\ \pm\sqrt{}}{2(\ )} =$ ___

2. $y = 3x^2 + 4x - 2$

   ___ - 4( )( )

   _____

   $\dfrac{-\ \pm\sqrt{}}{2(\ )} =$ ___

   $\dfrac{-\ \pm\sqrt{}}{2(\ )} =$ ___

3. $y = 2x^2 + 5x + 1$

   ___ - 4( )( )

   _____

   $\dfrac{-\ \pm\sqrt{}}{2(\ )} =$ ___

   $\dfrac{-\ \pm\sqrt{}}{2(\ )} =$ ___

Review 1. What does the quadratic formula do? _____
2. What formula starts the quadratic formula? _____
3. What is the formula for the Discrirminant? _____
4. Name 4 steps to solve a quadratic formula. _____

_____

5. $y = 4x^2 + 6x + 2$

   ___ - 4( )( )

   _____

   $\dfrac{-\ \pm\sqrt{}}{2(\ )} =$ ___

   $\dfrac{-\ \pm\sqrt{}}{2(\ )} =$ ___

6. $y = 3x^2 + 3x + 5$

   ___ - 4( )( )

   _____

   $\dfrac{-\ \pm\sqrt{}}{2(\ )} =$ ___

   $\dfrac{-\ \pm\sqrt{}}{2(\ )} =$ ___

Ch 3 Ls 3   Add and subtractract Cube Expressions.   23

_____ #1 #2 ____/ 8 #3 ____/ 6 R ____/ 9 T ____/ 23 _____
           Name   Checker

#1 1. What's the 1st step to factor Subtracted Cubes? _____
   2. How do cube roots make a "sandwich"? _____
   3. What do you get the middle term? _____
   4. If the cubed terms are added, how is it different? _____
   5. $a^3 + 8$ What's the 1st step this time? _____

#2 1. What's the 1st step to factor these? $a^3 - b^3$

   Find the sandwich 1st step.   (   )

   Find the sandwich 2nd step.   (   ) (   )
                                 (   ) (   )

   2. What's the 1st step to factor these cubes? $a^3 + 2^3$

   Find the sandwich 1st step.   (   )

   Find the sandwich 2nd step.   (   ) (   )
                                 (   ) (   )

   3. What's the 1st step now? $a^3 + 27$

   What happens next? _____

   Find the sandwich 1st step.   (   )

   Find the sandwich 2nd step.   (   ) (   )
                                 (   ) (   )

#3 Factor the cubed terms.   Calculator? yes   no

1. $a^3 - b^3$ $a^3 - 2^3$

   _____   _____
   _____   _____

2. $a^3 + b^3$ $a^3 + 2^3$

   _____   _____
   _____   _____

3. $a^3 - 8$ $a^3 + 27$

   _____   _____
   _____   _____

Review
1. What's the 1st step to factor Subtracted Cubes? _____
2. How do cube roots make a "sandwich"? _____
3. What do you get the middle term? _____
4. If the cubed terms are added, how is it different? _____
5. $a^3 + 8$ What's the 1st step this time? _____

6. $a^3 + \frac{1}{8}$ $a^3 - 27$

   _____   _____
   _____   _____

7. $a^3 + \frac{1}{27}$ $a^3 - \frac{1}{27}$

   _____   _____
   _____   _____

Review Problems   25

_____ #1 - #3 _____/ 11 Back _____/ 10 Total _____/ 21
             Name

1. Complete the Square _____
2. Quadratic Formula _____
3. Determinant _____
4. Subtracted Cubes _____
5. Added Cubes _____

#2   1. $x^2 + 8x + 3 = 0$ $x^2 + 6x + 2 = 0$

Use Complete
the Square.
          Divide: ___ Square: ___   Divide: ___ Square: ___

     2. $x^2 + 5x + 2 = 0$ $x^2 + 4x + 1 = 0$

          Divide: ___ Square: ___   Divide: ___ Square: ___

     3. $x^2 + 3x + 1 = 0$ $x^2 + 2x + 7 = 0$

          Divide: ___ Square: ___   Divide: ___ Square: ___

#6   1. $y = 2x^2 + 3x + 8$ $\dfrac{- \pm \sqrt{}}{2()} = \underline{}$ Calculator?

Use quadratic formulas. $\underline{} - 4()()$ yes   no

$\underline{}$ $\dfrac{- \pm \sqrt{}}{2()} = \underline{}$

2. $y = 3x^2 + 5x - 7$ $\dfrac{- \pm \sqrt{}}{2()} = \underline{}$

$\underline{} - 4()()$

$\underline{}$ $\dfrac{- \pm \sqrt{}}{2()} = \underline{}$

3. $y = 3x^2 + 7x - 1$ $\dfrac{- \pm \sqrt{}}{2()} = \underline{}$

$\underline{} - 4()()$

$\underline{}$ $\dfrac{- \pm \sqrt{}}{2()} = \underline{}$

4. $y = 2x^2 + 8x - 2$ $\dfrac{- \pm \sqrt{}}{2()} = \underline{}$

$\underline{} - 4()()$

$\underline{}$ $\dfrac{- \pm \sqrt{}}{2()} = \underline{}$

#7   1. $a^3 + \dfrac{1}{8}$ $a^3 - 27$

Factor these.

2. $a^3 + \dfrac{1}{27}$ $a^3 - \dfrac{1}{27}$

3. $a^3 + \dfrac{1}{64}$ $a^3 - 64$

Ch 4 Ls 1   Divide by a binomial.   27

_____ #1 #2 ____/ 7   #3 ____/ 10   R ____/ 4   T ____/ 21 _____
         Name   Checker

#1 1. What's the 1st step to divide by a binomial? _____
       2. What 2 steps solve each step for division? _____
       3. What do you do to make the 2nd? _____
       4. How do you make a remainder? _____

   #2 1. Divide the 1st step below. $x + 1 \overline{\smash{\big)}\, 2x^2 + 2x}$

              What's the remainder? $x + 1 \overline{\smash{\big)}\, 2x^2 + 2x}$

          2. Divide the 1st step below. $b + 3 \overline{\smash{\big)}\, 2b^2 + 6b}$

              What's the remainder? $b + 3 \overline{\smash{\big)}\, 2b^2 + 6b}$

          3. Solve it below.
             What's the 1st step? $a + 2 \overline{\smash{\big)}\, a^2 + 4a + 9}$

              Solve it here. $a + 2 \overline{\smash{\big)}\, a^2 + 4a + 9}$

1.

$b + 3 \overline{) 2b^2 + 6b}$ $x + 2 \overline{) 3x^2 + 6x}$ $3c + 1 \overline{) 2c^2 + 9c}$ Calculator? yes no

2.

$x + 2 \overline{) 2x^2 + 2x}$ $2b + 3 \overline{) 2b^2 + 2b}$ $a + 5 \overline{) 2a^2 + 2a}$

3.

$d + 2 \overline{) d^2 + 4d + 3}$ $2b + 3 \overline{) 2b^2 + 3b + 8}$

4.

$3x + 1 \overline{) 6x^2 + 2x + 7}$ $c + 4 \overline{) c^2 + 4c + 5}$

Review 1. What's the 1st step to divide by a binomial? _____   Calculator?
2. What 2 steps solve each step for division? _____   yes no
3. What do you do to make the 2nd? _____
4. How do you make a remainder? _____

_____

Ch 3 Ls 2 Divide with a missing term. 29

_____ #1 #2 ____/ 7 #3 ____/ 6 R ____/ 4 T____/ 17 _____
　　　Name　　　　　　　　　　　　　　　　　　　　　　　　　　　　　　　Checker

#1 1. How do you divide when there's a missing term? _____
　　　2. What key question starts to divide by a binomial? _____
　　　3. What 2 steps solves each step? _____
　　　4. How do you get a remainder? _____

#2 1. What does the problem divide?　　　$\dfrac{2x^2 - 4}{x - 3}$

　　　Divide the 1st step.

　　　2. Write it out and divide.　　　$\dfrac{4x^2 - 7}{x - 2}$

　　　Divide the 1st step.

　　　3. What does the problem divide?　　　$\dfrac{5x^2 - 3}{x - 1}$

　　　Divide the 1st step.

#3 1. Divide these quadratics.

$$x+2 \overline{\smash{)}\, 3x^3 + 5x^2 + 0x - 2}$$

$$c+3 \overline{\smash{)}\, 2c^3 + 0c^2 + 9c + 7}$$

Calculator? yes   no

2. $$2a-3 \overline{\smash{)}\, 4a^3 + 0a^2 + 5a + 1}$$

$$b+1 \overline{\smash{)}\, b^3 + 2b^2 + 0b + 3}$$

3. $$x+4 \overline{\smash{)}\, 4x^3 + 6x^2 + 0x - 5}$$

$$3c-5 \overline{\smash{)}\, 6c^3 + 0c^2 - 4c - 7}$$

Review 1. How do you divide when there's a missing term? _____
2. What key question starts to divide by a binomial? _____
3. What 2 steps solves each step? _____
4. How do you get a remainder? _____
_____

Calculator?   yes   no

Ch 3 Ls 3 Synthetic Division. 31

_____ #1 #2 ____/ 7 #3 ____/ 8 R ____/ 4 T____/ 19 _____
　　　　　Name　　　　　　　　　　　　　　　　　　　　　　　　　　　　　　　　Checker

#1 1. How is Synthetic Division different from box division? _____
_____

2. What's the 1st step to set up synthetic division? _____
3. What's the 1st step to divide it? _____
4. Where does the answer go? _____

#2 1. What does it divide by? Divide it. $b^2 + 6b + 8 \div b + 1$

$$\underline{|}\ 1\quad 6\quad 8$$

$$\overline{}$$

What does the answer mean? _____

$$\overline{}$$

2. What does it divide by? Divide it. $c^2 + 2c + 4 \div c + 3$

$$\underline{|}\ 1\quad 2\quad 4$$

$$\overline{}$$

What does the answer mean? _____

$$\overline{}$$

3. What does it divide by? Divide it. $x^2 + 5x + 3 \div x - 2$

$$\underline{|}\ 1\quad 5\quad 3$$

$$\overline{}$$

What does the answer mean? _____

$$\overline{}$$

#3 Divide these using Synthetic Division.   Calculator? yes no

1. $b^2 + 6b + 8 \div b - 3$ $2b^2 + 5b + 3 \div b + 1$

   $\underline{3|}\ 1\ \ -6\ \ \ 9$ $\underline{-1|}\ 2\ \ \ 5\ \ \ 3$

   _____   _____
   Answer

2. $4b^2 + 5b + 5 \div b - 2$ $3b^2 + 7b - 9 \div b + 3$

   $\underline{2|}\ 4\ \ \ 5\ \ \ 5$ $\underline{-3|}\ 3\ \ \ 7\ \ -9$

   _____   _____

3. $5b^2 + 6b + 3 \div b + 1$ $2b^2 - 5b + 3 \div b + 1$

   $\underline{-1|}\ 5\ \ \ 6\ \ \ 3$ $\underline{-3|}\ 2\ \ -5\ \ \ 3$

   _____   _____

4. $3b^2 + 4b - 8 \div b - 2$ $4b^2 + 6b - 8 \div b - 4$

   $\underline{2|}\ 1\ \ \ 4\ \ -8$ $\underline{4|}\ 1\ \ -6\ \ \ 9$

   _____   _____

Review 1. How is Synthetic Division different from box division? _____   Calculator? yes no
_____
2. What's the 1st step to set up synthetic division? _____
3. What's the 1st step to divide it? _____
4. Where does the answer go? _____
5. What does the 1st place in the answer stand for? _____

Ch 3 Ls 4 Nonstandard Synthetic Division 33

_____ #1 #2 ____ / 7 #3 ____ / 4 R ____ / 8 T ____ / 19 _____
            Name   Checker

#1 1. If a term is missing how does synthetic division divide it? _____
_____

2. What does standard form mean? _____
3. How does it make standard form? _____
4. How does Synthetic Division make a perfect C term? _____

#2 1. What does it divide by? Divide it. $2b^2 + 6b + 8 \div 2b + 1$

$$\underline{|}\; 1 \quad 3 \quad 4$$

$$\overline{}$$

What does the answer mean?

$$\overline{}$$

2. What does it divide by? Divide it. $3b^2 + 6b + 4.5 \div 3b + 2$

$$\underline{|}\; 1 \quad 2 \quad 1.5$$

$$\overline{}$$

What does the answer mean?

$$\overline{}$$

3. What does it divide by? Divide it. $b^2 + 5b + 4 \div 2b + 1$

$$\underline{|}\; 0.5 \quad 2.5 \quad 2$$

$$\overline{}$$

What does the answer mean?

$$\overline{}$$

#3 Solve these with missing terms.   Calculator? yes no

1. $4a^2 + 6a + 5 \div 2a - 2$ $6b^2 + 6b + 2 \div 3b + 1$

   ⌋ _____   ⌋ _____

2. $2c^2 + 3c + 8 \div 2c - 3$ $9x^2 + 6x + 7 \div 3x + 2$

   ⌋ _____   ⌋ _____

Review 1. If a term is missing how does synthetic division divide it? _____   Calculator? yes no
_____

2. What does standard form mean? _____

3. How does it make standard form? _____

4. How does Synthetic Division make a perfect C term? _____
_____

Find a Perfect C Term

5. $a^2 + 6a + ? \div a + 3$ $2b^2 + 4b - 6 \div b + 1$

   ⌋ 1   6 ____   ⌋ 2   4 ____

   _____   _____

6. $5c^2 - 10c + 8 \div c + 2$ $3x^2 - 6x + 5 \div x + 1$

   ⌋ 5 -10 ____   ⌋ 3 -6 ____

   _____   _____

Review Problems

_____ #1 - #3 ____ / 13   Back ____ / 12   Total ____ / 25
Name

1. Box Remainder _____
2. Missing Term _____
3. Synthetic Division _____
4. Remainder _____

#3 Divide these quadratics.

1.
$$x + 2 \overline{)\, 2x^2 + 2x\,} \qquad 2b + 3 \overline{)\, 2b^2 + 2b\,} \qquad a + 5 \overline{)\, 2a^2 + 2a\,}$$

2.
$$d + 2 \overline{)\, d^2 + 4d + 3\,} \qquad 2b + 3 \overline{)\, 2b^2 + 3b + 8\,}$$

3.
$$x + 2 \overline{)\, 3x^3 + 5x^2 + 0x - 2\,} \qquad c + 3 \overline{)\, 2c^3 + 0c^2 + 9c + 7\,}$$

4.
$$2a - 3 \overline{)\, 4a^3 + 0a^2 + 5a + 1\,} \qquad b + 1 \overline{)\, b^3 + 2b^2 + 0b + 3\,}$$

#3 Divide these using Synthetic Division.

Calculator? yes no

1. $a^2 + 6a + 8 \div b - 3$ $2b^2 + 5b + 3 \div b + 1$

    $\underline{3|}$ 1   6   8 $\underline{-1|}$ 2   5   3

    _____   _____

2. $4b^2 + 5b + 5 \div b - 2$ $3x^2 + 7x - 9 \div x + 3$

    $\underline{2|}$ 4   5   5 $\underline{-3|}$ 3   7   -9

    _____   _____

3. $5c^2 + 6c + 3 \div c + 1$ $2b^2 - 5b + 3 \div b + 1$

    $\underline{-1|}$ 5   6   3 $\underline{-3|}$ 2   -5   3

    _____   _____

#3 Divide using Synthetic Division.

1. $a^2 + 6a + 7 \div 2a + 3$ $2b^2 + 4b - 6 \div 2b + 1$

    $\rfloor$ $\rfloor$

    _____   _____

2. $8c^2 - 10c + 8 \div 4c + 2$ $3x^2 - 6x + 9 \div 3x + 1$

    $\rfloor$ $\rfloor$

    _____   _____

3. $4a^2 + 6a + 5 \div 2a - 2$ $6b^2 + 6b + 2 \div 3b + 1$

    $\rfloor$ $\rfloor$

    _____   _____

Ch 4 Ls 1   Fundamental Theroem of Algebra   35

_____ #1 #2 ____/ 9   #3 ____/ 10   R ____/ 5   T ____/ 24   _____
     Name   Checker

#1   1. What is the Fundamental Theorem of Algebra? _____

2. Name 3 steps to find X intercepts. _____

_____

3. What does the biggest exponent show about the x intercepts? _____

_____

4. Why is it "possible x intercepts"? _____

5. How does biggest exponent count turning points? _____

#2   1. How many possible X intercepts are there? $3x^2 + x - 1 = y$

How many possible turning points? _____

_____

2. How many possible X intercepts are there? $7x^4 + x^2 + 5 = y$

How many possible turning points? _____

_____

3. How many possible X intercepts are there? $2x^5 + 7x^3 - 8 = y$

How many possible turning points? _____

_____

4. How many possible X intercepts are there? $9x^7 + 4x^4 - 3 = y$

How many possible turning points? _____

_____

#3 How many possible X intercepts and turning points are there?   Calculator? yes   no

1. $5x^7 + 3x^4 - 1 = y$ $3x^2 + x - 1 = y$

X Intercepts 0 1 2 3 4 5 6 7   0 1 2 3 4 5 6 7
Turning Points 0 1 2 3 4 5 6 7   0 1 2 3 4 5 6 7

2. $7x^4 + x^2 + 5 = y$ $2x^5 + 7x^3 - 8 = y$

X Intercepts 0 1 2 3 4 5 6 7   0 1 2 3 4 5 6 7
Turning Points 0 1 2 3 4 5 6 7   0 1 2 3 4 5 6 7

3. $9x^6 + 2x^3 - 3 = y$ $3x^3 + x - 1 = y$

X Intercepts 0 1 2 3 4 5 6 7   0 1 2 3 4 5 6 7
Turning Points 0 1 2 3 4 5 6 7   0 1 2 3 4 5 6 7

4. $7x^9 + x^2 + 5 = y$ $2x^6 + 7x^3 - 8 = y$

X Intercepts 0 1 2 3 4 5 6 7   0 1 2 3 4 5 6 7
Turning Points 0 1 2 3 4 5 6 7   0 1 2 3 4 5 6 7

5. $7x^5 + x^2 + 5 = y$ $2x^4 + 7x^3 - 8 = y$

X Intercepts 0 1 2 3 4 5 6 7   0 1 2 3 4 5 6 7
Turning Points 0 1 2 3 4 5 6 7   0 1 2 3 4 5 6 7

Review 1. What is the Fundamental Theorem of Algebra? _____   Calculator?
        2. Name 3 steps to find X intercepts. _____   yes   no
        _____
        3. What does the biggest exponent show about the x intercepts? _____
        _____
        4. Why is it "possible x intercepts"? _____
        5. How does biggest exponent count turning points? _____

Ch 4 Ls 2   Remainder and Factor Theorems   39

_____ #1 #2 ____ / 9 #3 ____ / 6   R ____ / 7   T ____ / 22 _____
              Name   Checker

#1 1. What does the Factor Theorem do? _____
2. What do you do to divide an equation by 1? _____
3. How does the Factor Theorem work? _____
4. What's a Depressed Polynomial? _____
5. Is 2, 3 a point on this graph? Solve the point 1st. $y = x^2 + 4x + 2$

What did you find? _____

Solve it with Remainder Theorem.
What binomial does it divide by? _____

$\underline{}|\;1\quad 4\quad 2$

_____

What is the other binomial?

_____

#2 1. What does the Remainder Theorem do? _____
2. Name the 1st way to find a point. _____
3. Name 3 ways to solve equations. _____
_____
4. Is 4, 5 a point on this graph? Solve the point 1st. $y = x^2 + 3x + 5$

What did you find? _____

Solve it with Remainder Theorem.
What binomial does it divide by? _____

$\underline{}|\;1\quad 3\quad 5$

_____

What is the other binomial?

_____

#3 Find the factor or remainder theorem.   Calculator?
                                                                                                                        yes   no

1. Is x - 1 a factor of this equation? $2x^2 - 6x - 5 \div x - 1$
   Set up the synthetic division.

                                                                        ___| 2 -6 -5

                                                                                 _____

2. What does the answer mean? _____

3. What does it divide by? Divide it. $4a^2 + 6a + 8 \div a + 2$

                                                                      ___| 4 6 8

                                                            _____

4. What does the answer mean? _____

5. What does it divide by? Divide it. $9b^2 + 6b - 7 \div b + 3$

                                                                    ___| 9 6 7

                                                            _____

6. What does the answer mean? _____

Review 1. What does the Factor Theorem do? _____   Calculator?
          2. What do you do to divide an equation by 1? _____   yes   no
          3. How does the Factor Theorem work? _____
          4. What's a Depressed Polynomial? _____
          5. What does the Remainder Theorem do? _____
          6. Name the 1st way to find a point. _____
          7. Name 3 ways to solve equations. _____

              _____

Ch 4 Ls 3 Make a graph from binomials. 41

_____ #1 #2 ____/ 7 #3 ____/ 4 R ____/ 3 T____/ 14 _____
Name   Checker

#1 1. What do binomials show about a graph? _____

2. What does the equation show? _____

3. How do X intercepts find graph directions? _____

#2 1. What are the zeros for this equation? $(x + 3)(x + 1)(x - 1) = y$

What are the signs for it? _____ _____ _____

Make a graph. What does it look like? _____ _____ _____

2. What are the zeros for this equation? $(x + 5)(x - 1)(x - 7) = y$

What are the signs for it? _____ _____ _____

Make a graph. What does it look like? _____ _____ _____

#3 What are the zeros for each equation? Graph them.　　　　Calculator?
　　　　　　　　　　　　　　　　　　　　　　　　　　　　　　　yes　no

1. (x + 5)(x + 2)(x - 1) = y　　　　(x + 8)(x + 3)(x - 2) = y

_____   _____   _____　　_____   _____   _____

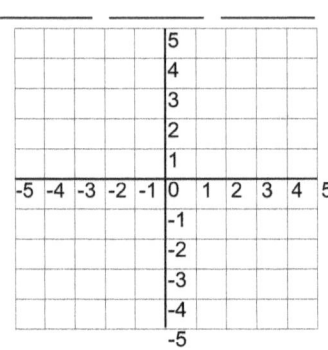　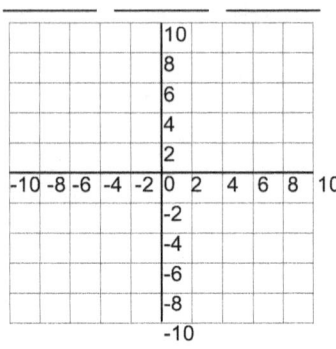

2. (x + 2)(x - 1)(x - 4) = y　　　　(x + 7)(x - 1)(x - 9) = y

_____   _____   _____　　_____   _____   _____

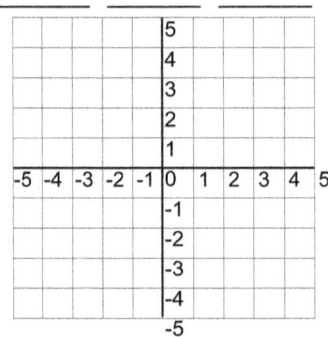

Review　1. What do binomials show about a graph? _____　Calculator?
　　　　2. What does the equation show? _____　yes　no
　　　　3. How do X intercepts find graph directions? _____

Review Problems 43

_____ #1 - #3 _____ / 8 Back _____ / 9 Total _____ / 17
         Name

1. Fundamental Theorem of Algebra _____

2. Remainder Theorem _____

3. Factor Theorem _____

4. Descartes Rule _____

#2 Find the zeros for equation. Graph.

1. What are the zeros for this equation? $(x + 4)(x + 1)(x - 2) = y$

    What are the signs for it?   _____   _____   _____

    Find the equation.   _____   _____   _____

Make a graph. What does it look like?   _____

#3 Find the factor or remainder theorem.

1. Is x - 1 a factor of this equation? $x^2 - 6x - 5 \div x - 1$
   Set up the synthetic division.

   ___| 1   - 6   - 5

   _____

2. What does it divide by? Divide it. $4a^2 + 6a + 8 \div a - 2$

   ___| 4   6   8

   _____

44

3. Is x - 4 a factor of this equation?
Set up the synthetic division.

$x^2 - 8x - 4 \div x - 4$

$\underline{|}\ 1\ \ -8\ \ -4$

_____

#3 What are the zeros for each equation? Graph them.

1. $(x + 5)(x + 2)(x - 1) = y$ $(x + 8)(x + 3)(x - 2) = y$ Calculator?
                                                                          yes   no

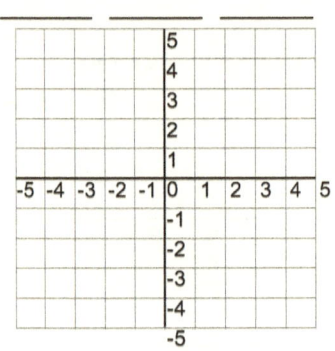 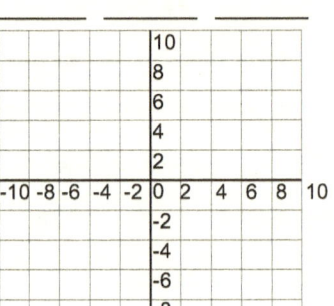

2. $(x + 2)(x - 1)(x - 4) = y$ $(x + 7)(x - 1)(x - 9) = y$

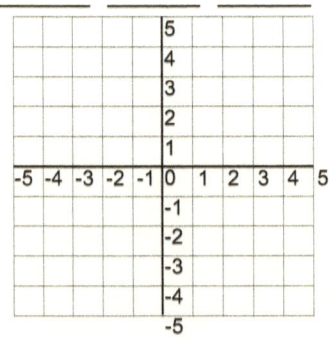 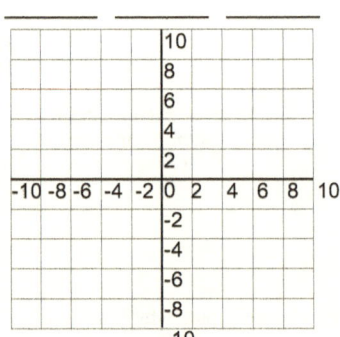

Ch 6 Ls 1   Begin Asymptotes   45

_____ #1 #2 ____/ 7   #3 ____/ 8   R ____/ 5   T____/ 20   _____
         Name   Checker

#1 1. What kind of equation makes an asymptote? _____

2. What numbers make a positive horizontal asymptote? _____

3. What happens when X is between 0 and 1? _____

4. Which line is the asymptote on a graph? _____

5. Name 2 kinds of asymptotes. _____

#2   1. Solve for x is 0.5, 2 and 5. $y = \dfrac{1}{x}$

0.5, ____   2, ____   5, ____   Graph both sides and sketch the asymptotes.

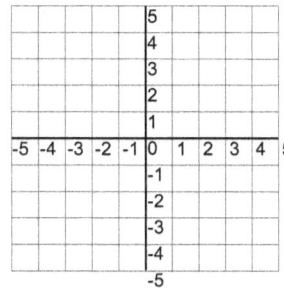

2. Solve for x is 0.5, 2 and 5. $y = \dfrac{2}{x} + 1$

0.5, ____   2, ____   5, ____   Graph both sides and sketch the asymptotes.

#3 Graph these.

1. $y = \dfrac{1}{x}$

If x is 2, then ___, ___

If x is 0.5, then ___, ___

Denominator _____

Solve for 2 points and denominator. Sketch the graph.

$y = \dfrac{1}{x} + 2$

If x is 2, then ___, ___

If x is 0.5, then ___, ___

Denominator _____

Calculator? yes no

2. $y = \dfrac{1}{x} - 3$

If x is 2, then ___, ___

If x is 0.5, then ___, ___

Denominator _____

Solve for 2 points and denominator. Sketch the graph.

$y = \dfrac{1}{x} + 4$

If x is 2, then ___, ___

If x is 0.5, then ___, ___

Denominator _____

Review
1. What kind of equation makes an asymptote? _____
2. What numbers make a positive horizontal asymptote? _____
3. What happens when X is between 0 and 1? _____
4. Which line is the asymptote on a graph? _____
5. Name 2 kinds of asymptotes. _____

Calculator? yes no

Ch 5 Ls 2 What makes asymptotes move? 47

_____ #1 #2 ____/ 6   #3 ____/ 8   R ____/ 4   T ____/ 18 _____
Name   Checker

#1 1. What makes an asymptote move right? _____
   2. What makes an asymptote move left? _____
   3. How can you solve to find where the asymptote is? _____
   4. How does X numerator change it? _____

#2 1. Solve for - 0.5, 0, and 5. $y = \dfrac{1}{x + 3}$

Graph both sides and sketch the asymptotes.   - 0.5, ____   0, ____   5, ____

2. Decide 3 points to solve for. $y = \dfrac{x}{x - 2}$

Graph both sides and sketch the asymptotes.   ____, ____   ____, ____   ____, ____

#3 Solve these.

1. $y = \dfrac{1}{x - 2}$

If x is 2, then ___, ___

If x is 0.5, then ___, ___

Denominator _____

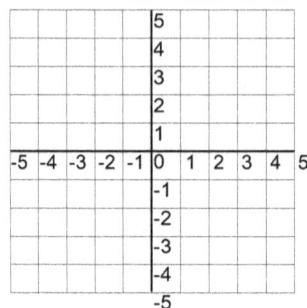

$y = \dfrac{x}{x - 1}$ Calculator? yes no

If x is 2, then ___, ___

If x is 0.5, then ___, ___

Denominator _____

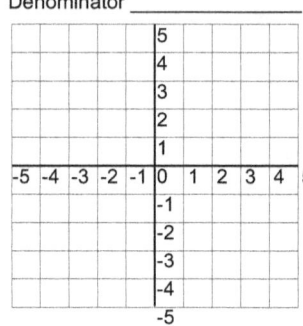

2. $y = \dfrac{1}{x + 3}$

If x is 2, then ___, ___

If x is 0.5, then ___, ___

Denominator _____

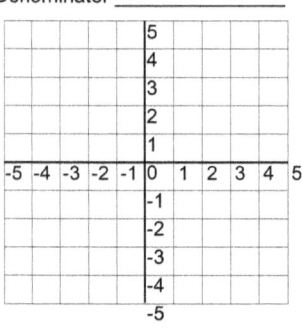

$y = \dfrac{x}{x + 4}$

If x is 2, then ___, ___

If x is 0.5, then ___, ___

Denominator _____

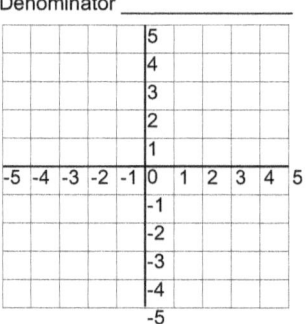

Review
1. What makes an asymptote move right? _____
2. What makes an asymptote move left? _____
3. How can you solve to find where the asymptote is? _____
4. How does X numerator change it? _____

Calculator? yes no

Ch 5 Ls 3   Binomial Numerators   49

_____ #1 #2 ____/ 8 #3 ____/ 8 R ____/ 3 T ____/ 19 _____
              Name   Checker

#1 1. What does a binomial numerator control? _____

2. What happens if it adds 1? _____

3. What happens if it subtracts 1? _____

4. Decide 3 points to solve for. $y = \dfrac{x - 2}{x + 2}$

Graph both sides and sketch the asymptotes.   ___,___   ___,___   ___,___

#2 1. How do 2 binomials make a graph? _____

2. How do you find the vertex between the asymptotes? _____

3. You find x. How to find the y part? _____

4. Where are the asymptotes? $y = \dfrac{1}{(x + 1)(x - 2)}$

x + 1 = 0   x = ____   x - 2 = 0   x = ____   Find x is 0, - 0.5, 1.5 and 4.
                                                      What does the graph look like?

If x is 0, ____, ____   x is - 2, ____, ____

If x is 3, ____, ____   x is - 0.5, ____, ____

Discontinuity _____

#3 1. $y = \dfrac{x-3}{x+1}$ $y = \dfrac{x-2}{x-4}$ Calculator? yes   no

If x is 2, then ___, ___   If x is 2, then ___, ___

If x is 0.5, then ___, ___   If x is 0.5, then ___, ___

Denominator _____   Denominator _____

Solve for 2 points.
Find a denominator.

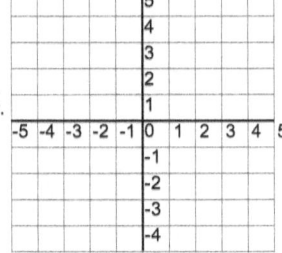 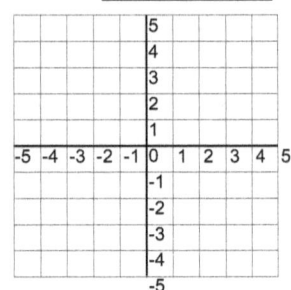

2. $y = \dfrac{1}{(x-1)(x+4)}$ $y = \dfrac{1}{(x+3)(x-3)}$

If x is -1, ___, ___   If x is 2, ___, ___   If x is 1, ___, ___   If x is -2, ___, ___

If x is -4, ___, ___   x is 0.5, ___, ___   If x is 4, ___, ___   If x is 0, ___, ___

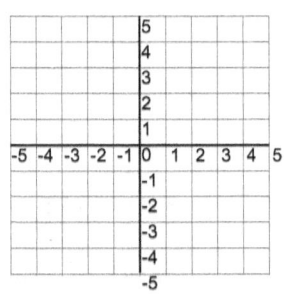 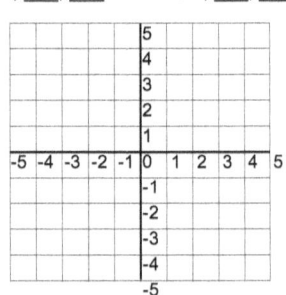

Review 1. What does a binomial numerator control? _____   Calculator?
2. What happens if it adds 1? _____   yes   no
3. How do 2 binomials make a graph? _____
4. How do you find the vertex between the asymptotes? _____
5. You find x. How to find the y part? _____

Review Problems 51

_____ #1 - #3 ____/ 7   Back ____/ 10   Total ____/ 17
            Name

1. **Asymptote** _____

2. **Horizontal Asymptotes** _____

3. **Vertical Asymptotes** _____

#2   1. $y = \dfrac{1}{x}$ $y = \dfrac{1}{x} + 2$

Graph these.
If x is 2, then ___, ___   If x is 2, then ___, ___

If x is 0.5, then ___, ___   If x is 0.5, then ___, ___

Denominator _____   Denominator _____

Solve for 2 points and denominator. Sketch the graph.

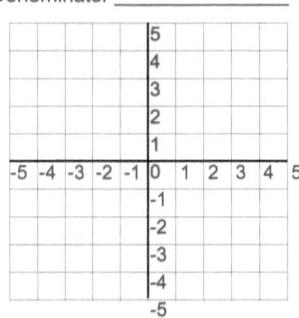 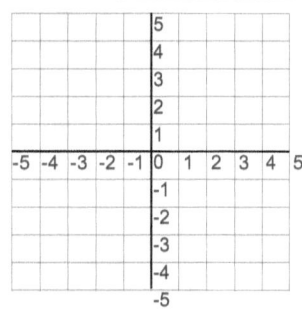

2. $y = \dfrac{1}{x - 5}$ $y = \dfrac{x}{x - 3}$

If x is 2, then ___, ___   If x is 2, then ___, ___

If x is 0.5, then ___, ___   If x is 0.5, then ___, ___

Denominator _____   Denominator _____

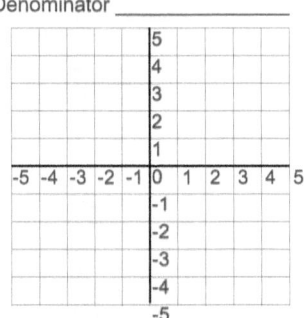 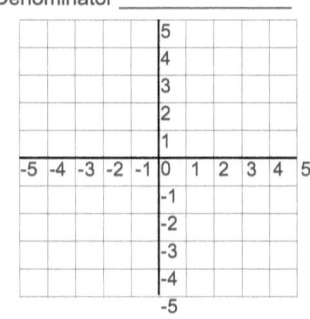

#3 1. $y = \dfrac{x - 5}{x + 2}$ $y = \dfrac{x - 3}{x - 6}$ Calculator?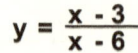

If x is 2, then ___, ___   If x is 2, then ___, ___

If x is 0.5, then ___, ___   If x is 0.5, then ___, ___

Denominator _____   Denominator _____

Solve for 2 points.
Find a denominator.

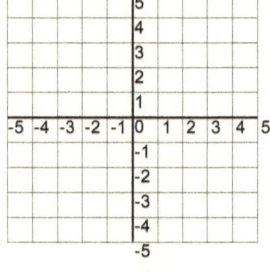   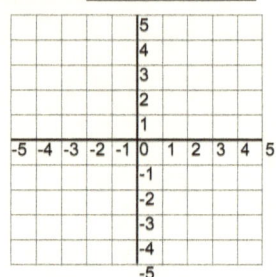

2. $y = \dfrac{1}{(x - 2)(x + 3)}$ $y = \dfrac{1}{(x + 3)(x - 3)}$

If x is -1, ___, ___   If x is 2, ___, ___   If x is 1, ___, ___   If x is -2, ___, ___

If x is -4, ___, ___   x is 0.5, ___, ___   If x is 4, ___, ___   If x is 0, ___, ___

Solve for 2 points.
Find a denominator.

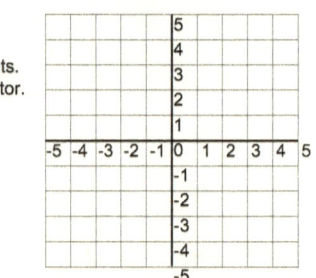   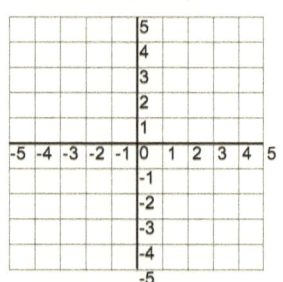

#4 1. $y = \dfrac{1}{(x - 5)(x + 1)}$ $y = \dfrac{1}{(x + 2)(x - 6)}$

Where does the
graph center
around?   _____   _____

_____   _____

Ch 7 Ls 1   Point Disconuity.   53

_____ #1 #2 ____ / 9 #3 ____ / 8 R ____ / 6 T ____ / 23 _____
Name   Checker

#1   1. What is the name for the hollow point? _____

2. Why does an equation make a discontinuity? _____

3. What rule graphs an equation that can be simplified? _____

4. How does this fraction simplify? $y = \dfrac{2x}{x}$

What's the discontinuity. Graph it. $y = $

Discontinuity _____

#2   1. How do you simplify a fraction with a binomial? _____

2. How do you find what's discontinued? _____

3. How do you graph a rational function that can be simplified? _____
_____

4. How does a graph show a point that is not included? _____

5. How does the fraction simplify? $y = \dfrac{(x + 1)}{(2x + 2)}$

Find the point of discontinuity.
What does the graph look like? $y = \dfrac{}{}$

Discontinuity _____

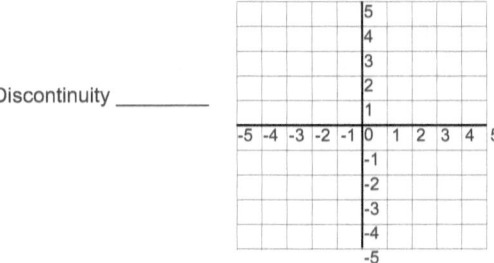

#3 Simplify. Find the point of discontinuity.

1. $y = \dfrac{3x}{x}$

   $y =$

   Discontinuity _____

   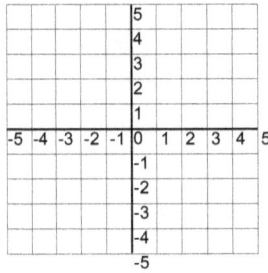

   $y = \dfrac{-4x}{x}$

   $y =$

   Discontinuity _____

   Calculator? yes no

   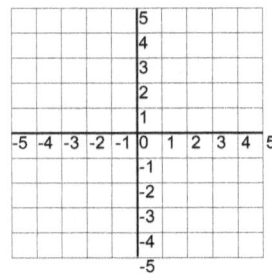

2. $y = \dfrac{(2x+2)}{(4x+8)}$

   $y =$ _____

   Discontinuity _____

   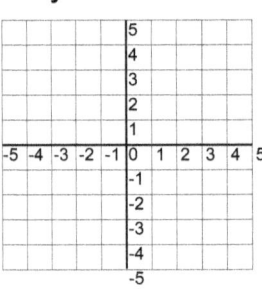

   $y = \dfrac{(x+2)}{(3x^2+6)}$

   $y =$ _____

   Discontinuity _____

   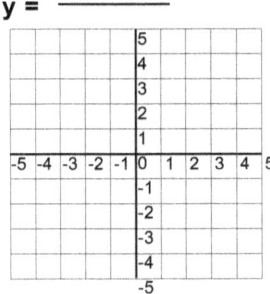

Review

1. What is the name for the hollow point? _____
2. Why does an equation make a discontinuity? _____
3. What rule graphs an equation that can be simplified? _____
4. How do you simplify a fraction with a binomial? _____
5. How do you find what's discontinued? _____
6. How do you graph a rational function that can be simplified? _____

_____

Calculator? yes no

Ch 7 Ls 2   Asymptote Discontinuities Pt 2   55

_____ #1 #2 ____/ 8   #3 ____/ 8   R ____/ 4   T ____/ 20   _____
            Name   Checker

#1 1. What is a point of discontinuity? _____

2. How do you graph this asymptote? $\dfrac{(x-1)(x-2)}{x-1}$ _____

_____

3. What's 1st to graph this asymptote? $\dfrac{(x+3)(x-1)}{x^2-9}$ _____

4. What's next? _____

#2 1. How does the fraction simplify? $y = \dfrac{(x+1)}{(2x+2)}$

Find the point of discontinuity.
What does the graph look like? $y = \dfrac{}{}$

Discontinuity _____

2. How does the fraction simplify? $y = \dfrac{(x+2)}{(x^2+4)}$

Find the point of discontinuity.
What does the graph look like? $y = \dfrac{}{}$

Discontinuity _____

56

#3 **1.** $y = \dfrac{(x + 3)}{(3x + 6)}$ $y = \dfrac{(x + 1)}{(x^2 + 4)}$ Calculator? yes no

Find the point of discontinuity.

$y = $ ――― $y = $ ―――

Discontinuity

―――

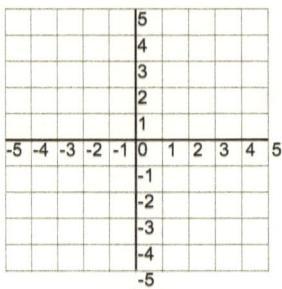

Discontinuity

―――

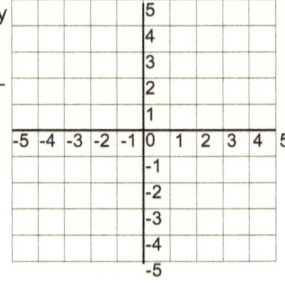

2. $y = \dfrac{(2x + 2)}{(4x + 8)}$ $y = \dfrac{(x + 2)}{(3x^2 + 6)}$

Find the point of discontinuity.

$y = $ ――― $y = $ ―――

Discontinuity

―――

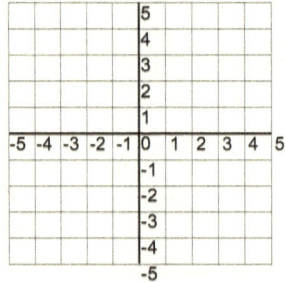

Discontinuity

―――

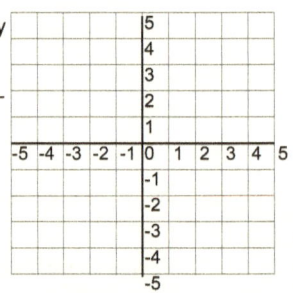

Review 1. What is a point of discontinuity? _____ Calculator? yes no

2. How do you graph this asymptote? $\dfrac{(x - 1)(x - 2)}{x - 1}$ _____

_____

3. What's 1st to graph this asymptote? $\dfrac{(x + 3)(x - 1)}{x^2 - 9}$ _____

4. What's next? _____

Ch 7 Ls 3   Asymptote story problems.   57

_____ #1 #2 ____/ 9 #3 ____/ 8 R ____/ 5 T ____/ 23 _____
          Name   Checker

#1 1. What's the 1st step to make an answer to the baseball problem? _____
_____

2. What's the 2nd step? _____
3. Name the 1st step for the T shirt problem. _____
4. Name the 2nd step. _____
5. What do you subtract? _____

#2 1. What's 1st?   Ojas has 3 hits out of 10 at bats. He wants to hit 500.
                                How many hits does he need in his next 10 at bats?

            How does it finish the fraction?   _____

                      Make a graph for it.   _____

_____ hits out of   20
                                     16
_____ at bats   Hits   12
                                     8
                                     4
                                     0
                           At bats   0   4   8 12 16 20

2. What's goes   A stdent concil wants to sell boxes of cookies at their school. Each
in the den-   box is Rs 150 snf there is a fee of 2000. Five student council mem-
ominator?   bers get a free box for try out. How many will they sell to break even?

          How mch is each box of cookeies?   _____

                      Finish the fraction.   _____

                                                _____

#3 1. **Amav has 5 hits out of 12 at bats. He wants to hit 500. How many hits does he need in his next 12 at bats?**

Calculator? yes no

He needs to hit _____ out of 12 to hit 500.

2. **Mitul has 2 hits out of 9 at bats. He wants to hit 300. How many hits does he need in his next 11 at bats?**

He needs to hit _____ out of 11 to hit 300.

3. **Zara sold boxes of cookies at her mom's factory. Each box is Rs 150 snf there is a fee of 300. She got 3 boxes for free. How many will she need to sell to break even?**

She needs to sell Rs _____ to break even

4. **Update: Zara sold alot, so she raised it to Rs 200 snf there is a fee of 500. She got 5 boxes for free. How many will she need to sell to break even?**

She needs to sell Rs _____ to break even

Review 1. What's the 1st step to make an answer to the baseball problem? _____

Calculator? yes no

2. What's the 2nd step? _____
3. Name the 1st step for the T shirt problem. _____
4. Name the 2nd step. _____
5. What do you subtract? _____

Review Problems 59

_____ #1 - #2 ____/ 6 Back ____/ 6 Total ____/ 12
Name

1. Discontinuity _____

2. Slant Asymptote _____

#2 Use asymptote laws to graph these.

1. $y = \dfrac{1}{x}$ $y = \dfrac{1}{x} + 2$

If x is 2, then ___ , ___   If x is 2, then ___ , ___

If x is 0.5, then ___ , ___   If x is 0.5, then ___ , ___

Denominator _____   Denominator _____

Solve for 2 points
and denominator.
Sketch the graph.

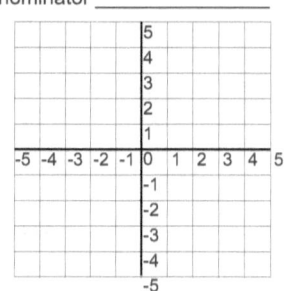   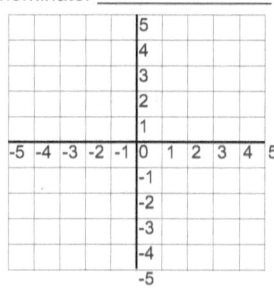

2. $y = \dfrac{1}{x - 2}$ $y = \dfrac{x}{x - 1}$

If x is 2, then ___ , ___   If x is 2, then ___ , ___

If x is 0.5, then ___ , ___   If x is 0.5, then ___ , ___

Denominator _____   Denominator _____

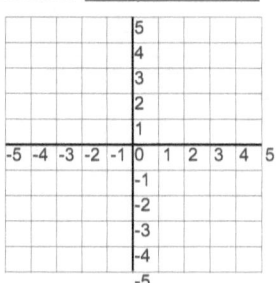   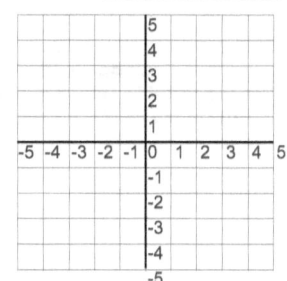

#3 1. Find the point of discontinuity.

$$y = \frac{(x+3)}{(3x+6)}$$

y = _____

Discontinuity

_____

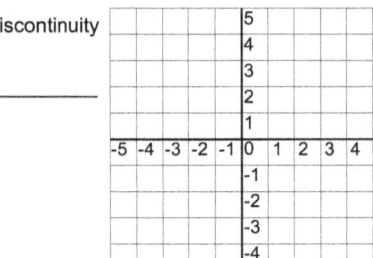

$$y = \frac{(x+1)}{(x^2+4)}$$

y = _____

Discontinuity

_____

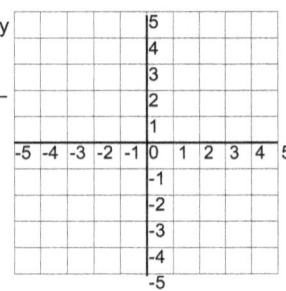

Calculator?   yes   no

#3

1. Hansh hit 5 goals out of 20 attempts. He wants to make 300. How many hits does he need in his next 10 attempts?

   He needs to hit _____ out of 10 to hit 300.

2. Kiaan has 7 hits out of 20 attemtpts in cricket. He wants to hit 500. How many hits does he need in his next 10 at bats?

   He needs to hit _____ out of 10 to hit 500.

3. The soccer club sold cookies at their school. Each box is Rs 300 snd there is a fee of 2000. They got 10 boxes for free. How many will they need to sell to break even?

   They needs to sell Rs _____ to break even.

Calculator?   yes   no

Ch 8 Ls 1   Direct and Indirect Variation   61

_____ #1 #2 ____/ 7 #3 ____/ 4 R ____/ 6 T____/ 17 _____
          Name   Checker

#1 1. What is Direct Variation? _____

   2. What is a Constant? _____

   3. What is an Inverse Variation? _____

   5 x 40 = 200   10 x 20 = 200

   4. Where is the konstant for these inverse equations? _____

   _____

#2 1. Make 2 equations.   It snowed 4 cm an hour for 3 hours. How much would it snow in 6 hours to equal the same amount?

                                 ___ x ___ = ___
   Is it a direct or indirect variation? Why?
                                 ___ x ___ = ___

         Direct   Inverse   Because _____.

   2. Make 2 equations.   The rocket used to studyears the moon's surface weighs 12,000 kg on the earth. Find how it weighs on the moon. (1/6 gravity of earth.

                                 ___ x _____ = _____
   Is it a direct or indirect variation? Why?
                                 ___ x _____ = _____

         Direct   Inverse   Because _____.

   3. Make 2 equations.   I ran a 400 meter race at 2 m/sec for a time of 200 sec (3 min 20 sec). The next day I ran the same race at 2.5 m/sec for 160 sec. (2:40)

                                 ___ x _____ = _____
   Is it a direct or indirect variation? Why?
                                 ___ x _____ = _____

         Direct   Inverse   Because _____.

#3 Decide which it is and why.   Calculator? yes no

1.   5 x 60 = 300   5 x 40 = 200
     10 x 30 = 300   10 x 40 = 400
Circle one.   Direct   Inverse   Direct   Inverse

_____   _____

2.   10 x 20 = 200   3 x 50 = 150
     40 x 20 = 800   5 x 30 = 150
     Direct   Inverse   Direct   Inverse

_____   _____

3.   5 x 40 = 200   4 x 50 = 200
     10 x 20 = 200   7 x 50 = 350
     Direct   Inverse   Direct   Inverse

_____   _____

Review   1. What is Direct Variation? _____   Calculator?
         2. What is a Constant? _____   yes   no
         3. What is an Inverse Variation? _____

            5 x 40 = 200   10 x 20 = 200
        4. Where is the konstant for these inverse equations? _____

5. TJ buys a dozen doughnuts for the office at Rs 15 ea
    or 15 doughnuts at 12 ea. Write a relation.   ____ x ____ = ____

                                                                                  ____ x ____ = ____

Is it a direct or indirect variation? Why? _____

6. With all his gear on, Neil Armstrong weighed 160 kg on
    earth, but weighed 25 kg on the moon. Ojas weighs 70   ____ x ____ = ____
    kg on earth. How much would he weigh on the moon?
                                                                                       ____ x ____ = ____

Is it a direct or indirect variation? Why? _____

Ch 8 Ls 2 Joint Variation   63

_____ #1 #2 ____ / 6 #3 ____ / 7 R ____ / 5 T ____ / 18 _____
           Name   Checker

#1 1. How is joint variation different from direct variation? _____
     2. How does the area of a triangle use 3 terms? _____
     3. What does direct, indirect, and joint variation do? Direct _____
       Indirect _____ Joint _____

     Make 2 equations, then decide which kind of variation it is.

#2 1. Make 2 equations.   We borrowed Rs 300,000. Either we had 12% loan for
                            2 years or a 14% loan for 3 years.

       Which kind of variation is it? Why?   ___ x _____ x ___ = ___
                                             ___ x _____ x ___ = ___

       Direct   Indirect   Joint   Because _____.

     2. Make 2 equations.   Ojas ran a 100 m race at 5 m/sec for 20 seconds, then
                            he ran again and he clocked 7 m/sec for 14.3 seconds.

       Which kind of variation is it? Why?   ___ x _____ = ___
                                             ___ x _____ = ___

       Direct   Indirect   Joint   Because _____.

     3. Make 2 equations.   Myra built a triangle that's 40 cm long and 20 cm high.
                            She wasn't happy, so she made one that's 45 cm long
                            and 18 cm high.

       Which kind of variation is it? Why?   ___ x _____ x ___ = ___
                                             ___ x _____ x ___ = ___

       Direct   Indirect   Joint   Because _____.

#3 Circle which it is and explain why.   Calculator?
                                                                   yes   no

1.   0.5 x 5 x 80 = 200   3 x 80 = 240
     0.5 x 10 x 40 = 200   10 x 24 = 240
     Direct   Indirect   Joint   Direct   Indirect   Joint

     _____   _____

2.   90 x 30 = 2700   3.1 x 10 x 60 = 1860
     27 x 100 = 2700   3.1 x 12 x 50 = 1860
     Direct   Indirect   Joint   Direct   Indirect   Joint

     _____   _____

3.   5 x 40 = 200   2 x 80 = 160
     10 x 40 = 400   5 x 80 = 400
     Direct   Indirect   Joint   Direct   Indirect   Joint

     _____   _____

Review 1. How is joint variation different from direct variation? _____   Calculator?
        2. How does the area of a triangle use 3 terms? _____   yes   no
        3. What does direct, indirect, and joint variation do? Direct _____
           Indirect _____   Joint _____

        4. Zara bought 6 zuchini at Rs 60 for Rs 360 and 10   ___ X ___ = ___
           bundles of carrot at the same price for Rs 600.
                    Make 2 equations.   ___ X ___ = ___

           Which kind of variation is it and why? _____

        5. Sana is buying tennis shoes. She could buy 3 for   ___ X ___ = ___
           Rs 500 each or she could buy 2 for Rs 750 each.
                    Make 2 equations.   ___ X ___ = ___

           Which kind of variation is it and why? _____

Ch 9 Ls 3   Multiply Fractions with Exponents   65

_____ #1 #2 ____ / 6   #3 ____ / 14   R ____ / 3   T ____ / 23 _____
              Name   Checker

#1 1. How do you multiply a fraction by a fraction? _____

2. What happens if the fractions have variables? _____

3. What happens if the fraction has trinomials? _____

#2 1. What's the 1st step to multiply these fractions? $\dfrac{4e^4}{5d^2} \cdot \dfrac{3d^4}{2e^5}$

Solve numbers 1st.
Finish it. What's the answer? $\dfrac{\quad}{\quad} \cdot \dfrac{\quad}{\quad}$

$\dfrac{\quad}{\quad}$

2. What's the 1st step to multiply these fractions? $\dfrac{2a^2 + 4}{2a^2 + a} \cdot \dfrac{a + 2}{a - 1}$

Factor the binomials.
Finish it. What's the answer? $\dfrac{\quad}{\quad} \cdot \dfrac{\quad}{\quad}$

$\dfrac{\quad}{\quad}$

3. What's the 1st step to multiply these fractions? $\dfrac{a^2 + 4a + 6}{a^2 + 4a + 6} \cdot \dfrac{a + 2}{a - 1}$

Factor the trinomials.
Finish it. What's the answer? $\dfrac{\quad}{\quad} \cdot \dfrac{\quad}{\quad}$

$\dfrac{\quad}{\quad}$

#3

1. $\dfrac{2a^4}{5a^2} \cdot \dfrac{2a^4}{3a^3} = \underline{\qquad}$ $\dfrac{3x^2}{9x^4} \cdot \dfrac{6x^4}{6x^5} = \underline{\qquad}$ Calculator? yes   no

2. $\dfrac{2b^7}{5b^2} \cdot \dfrac{3b^4}{2b^5} = \underline{\qquad}$ $\dfrac{6y^3}{5y^2} \cdot \dfrac{3y^4}{3y^5} = \underline{\qquad}$

3. $\dfrac{8d^4}{6c^2} \cdot \dfrac{2c^4}{4d^3} = \underline{\qquad}$ $\dfrac{20z^2}{9x^4} \cdot \dfrac{6x^5}{6z^5} = \underline{\qquad}$

4. $\dfrac{2d^5}{5d^2} \cdot \dfrac{15d^4}{4d^7} = \underline{\qquad}$ $\dfrac{4g^4}{9e^2} \cdot \dfrac{6e^4}{8g^5} = \underline{\qquad}$

5. $\dfrac{6a^2 + 4}{2a^2 + a} \cdot \dfrac{a - 1}{3a^2 + 2} = \underline{\qquad}$

6. $\dfrac{2b - 6}{4b^2 + 2b} \cdot \dfrac{b^2 + 1}{b - 3} = \underline{\qquad}$

7. $\dfrac{3c^2 + 3}{2c^2 + c} \cdot \dfrac{c^2 + c}{c^2 + 1} = \underline{\qquad}$

8. $\dfrac{5}{x^2 + 4x + 3} \cdot \dfrac{x + 3}{x - 1} = \underline{\qquad}$

9. $\dfrac{y + 4}{2y^2 + 3y - 2} \cdot \dfrac{y + 2}{y - 3} = \underline{\qquad}$

10. $\dfrac{2z^2 - 4}{4z^2 + 4z - 8} \cdot \dfrac{z + 2}{z - 2} = \underline{\qquad}$

Review
1. How do you multiply a fraction by a fraction? _____   Calculator? yes   no
2. What happens if the ractions have variables? _____
3. What happens if the fraction has trinomials? _____

Ch 9 Ls 4 Divide and Complex Fractions   67

_____ #1 #2 ____/ 8 #3 ____/ 8 R ____/ 8 T ____/ 24 _____
            Name   Checker

#1 1. How do you divide with fractions? _____
2. What happens if the fractions have variables? _____
3. What happens if the fraction has trinomials? _____

4. What's the 1st step to multiply these fractions?

$$\frac{4e^4}{5d^2} \div \frac{3d^4}{2e^5}$$

Solve numbers 1st.
Finish it. What's the answer?

$$\frac{4e^4}{5d^2} \bullet \underline{}$$

$$\frac{\underline{}}{\underline{}}$$

5. What's the 1st step to multiply these fractions?

$$\frac{3}{a^2 + a - 3} \div \frac{a - 1}{a + 2}$$

Factor the trinomials.
Finish it. What's the answer?

$$\frac{3}{\underline{}} \bullet \frac{\underline{}}{\underline{}}$$

$$\underline{}$$

#2 1. What is a partial fraction? _____
2. How do partial fractions get same denominators? _____
3. How can a trinomial get same denominators? _____
4. How do you get same denominators?

$$\frac{1}{2x - 1} + \frac{1}{8x - 2}$$

What's next?

$$\frac{\underline{}}{\underline{}} \bullet \frac{1}{2x - 1} + \frac{1}{8x - 2}$$

Finish it. What is it added?

$$\frac{\underline{}}{\underline{}} + \frac{1}{8x - 2}$$

$$\frac{\underline{}}{8x - 2}$$

#3 Divide these.

1. $\dfrac{2a^4}{5b^2} \div \dfrac{3a^3}{2b^4} =$ $\dfrac{3x^2}{9y^4} \div \dfrac{6x^5}{6y^3} =$ Calculator? yes no

$\dfrac{2a^4}{5b^2} \cdot \dfrac{}{} = \dfrac{}{}$ $\dfrac{3x^2}{9y^4} \cdot \dfrac{}{} = \dfrac{}{}$

2. $\dfrac{5c^4}{10d^2} \div \dfrac{5c^3}{2d} =$ $\dfrac{4x^2}{5x^4} \div \dfrac{6x^4}{15x^5} =$

$\dfrac{5c^4}{10d^2} \cdot \dfrac{}{} = \dfrac{}{}$ $\dfrac{4x^2}{5x^4} \cdot \dfrac{}{} = \dfrac{}{}$

3. $\dfrac{3a^2+4}{2b^3+b} \div \dfrac{a+2}{6b-1} =$ $\dfrac{2b^2+4}{7x^2+3} \div \dfrac{b+2}{14x-6} =$

$\dfrac{3a^2+4}{2b^3+b} \cdot \dfrac{}{} = \dfrac{}{}$ $\dfrac{2b^2+4}{7x^2+3} \cdot \dfrac{}{} = \dfrac{}{}$

4. $\dfrac{3c^2+12}{2y^2-4} \div \dfrac{c+4}{y^3-1} =$ $\dfrac{-3x^2+9}{2x^2+4} \div \dfrac{-x+3}{x+2} =$

$\dfrac{3c^2+12}{2y^2-4} \cdot \dfrac{}{} = \dfrac{}{}$ $\dfrac{-3x^2+9}{2x^2+4} \cdot \dfrac{}{} = \dfrac{}{}$

Review
1. How do you divide with fractions? _____   Calculator? yes no
2. What happens if the fractions have variables? _____
3. What happens if the fraction has trinomials? _____
4. What is a partial fraction? _____
5. How do partial fractions get same denominators? _____
6. How can a trinomial get same denominators? _____

Partial fractions

7. $\dfrac{1}{4x-1} + \dfrac{1}{16x-2}$ $\dfrac{1}{9x-1} + \dfrac{1}{36x-18}$

$\underline{} \cdot \dfrac{1}{4x-1} + \dfrac{1}{16x-2}$ $\underline{} \cdot \dfrac{1}{9x-1} + \dfrac{1}{36x-18}$

$\underline{}$ $\underline{}$

Review Problems   69

_____ #1 - #3 ____ / 9 Back ____ / 22 Total ____ / 31
           Name

1. Direct Variation _____

2. Indirect Variation _____

3. Constant _____

4. Joint Variation _____

#2   Is it direct, inverse, or joint variation and why.

1. Ojas has a construction company. He bought 30 sheets of plywood for Rs 250 and 30 mirrors for them at Rs 400 each. List the equations. Which variation is it?

   _____ x _____ = _____

   _____ x _____ = _____

   Direct   Inverse   Joint

2. Amav found the car of his dreams!! Now he has to find a way to pay for it. He has 2 different bank loans, 1 for Rs 1200 for 8% at 24 months and another one for 12% for 48 months. Which variation is it?

   ___x___ x _____ = _____

   ___x___ x _____ = _____

   Direct   Inverse   Joint

3. Adah bought 6 cans of soda for Rs 25 each or 8 pieces of candy for the same price. Which variation is it?

   _____ x _____ = _____

   _____ x _____ = _____

   Direct   Inverse   Joint

4. Naveed found 2 houses he likes. He can get #1 for 6% at 10 years for Rs 5000 per month or he could get #2 for 6% at 12 years for Rs 5000 per month. Which variation is it?

   _____ x _____ = _____

   _____ x _____ = _____

   Direct   Inverse   Joint

5. Keya is buying concrete bricks for her basement. She could get 100 for Rs 15 each or 100 for Rs 12 each (but she has to carry them herself). Which variation is it?

   _____ x _____ = _____

   _____ x _____ = _____

   Direct   Inverse   Joint

#3 1. $\dfrac{2a^4}{5a^2} \cdot \dfrac{2a^4}{3a^3} = $ _____ 　　$\dfrac{3x^2}{9x^4} \cdot \dfrac{6x^4}{6x^5} = $ _____ 　Calculator? yes　no

2. $\dfrac{2b^7}{5b^2} \cdot \dfrac{3b^4}{2b^5} = $ _____ 　　$\dfrac{6y^3}{5y^2} \cdot \dfrac{3y^4}{3y^5} = $ _____

3. $\dfrac{8d^4}{6c^2} \cdot \dfrac{2c^4}{4d^3} = $ _____ 　　$\dfrac{20z^2}{9x^4} \cdot \dfrac{6x^5}{6z^5} = $ _____

4. $\dfrac{2d^5}{5d^2} \cdot \dfrac{15d^4}{4d^7} = $ _____ 　　$\dfrac{4g^4}{9e^2} \cdot \dfrac{6e^4}{8g^5} = $ _____

#4 1. $\dfrac{2a^4}{5b^2} \div \dfrac{2b^4}{3a^3} = $ 　　　$\dfrac{3x^2}{9y^4} \div \dfrac{6y^3}{6x^5} = $ 　Calculator? yes　no

Divide these.

$\dfrac{2a^4}{5b^2} \cdot $ _____ $ = $ _____ 　　$\dfrac{3x^2}{9y^4} \cdot $ _____ $ = $ _____

2. $\dfrac{5c^4}{10d^2} \div \dfrac{2d}{5c^3} = $ 　　　$\dfrac{4x^2}{5x^4} \div \dfrac{6x^4}{15x^5} = $

$\dfrac{5c^4}{10d^2} \cdot $ _____ $ = $ _____ 　　$\dfrac{4x^2}{5x^4} \cdot $ _____ $ = $ _____

#5 1. $\dfrac{1}{4x-1} + \dfrac{1}{8x-2} = $ ___ $\cdot \dfrac{1}{4x-1} + \dfrac{1}{8x-2} = $ _____ 　Calculator? yes　no

Find the partial fractions.

2. $\dfrac{1}{3x-1} + \dfrac{1}{6x-3} = $ ___ $\cdot \dfrac{1}{3x-1} + \dfrac{1}{6x-3} = $ _____

3. $\dfrac{1}{3x-1} + \dfrac{1}{9x-3} = $ ___ $\cdot \dfrac{1}{3x-1} + \dfrac{1}{9x-3} = $ _____

4. $\dfrac{1}{6x-2} + \dfrac{1}{12x-4} = $ ___ $\cdot \dfrac{1}{6x-2} + \dfrac{1}{12x-4} = $ _____

5. $\dfrac{1}{5x-2} + \dfrac{1}{15x-6} = $ ___ $\cdot \dfrac{1}{5x-2} + \dfrac{1}{15x-6} = $ _____

Ch 9 Ls 1 Add with binomial denominators. 71

_____ #1 #2 ____/ 8 #3 ____/ 6 R ____/ 5 T____/ 19 _____
         Name   Checker

#1 1. What's the 1st step with an equation with fractions? _____
       2. After writing the numbers over them, what's next? _____
       3. What happens to what's left over? _____
       4. If there are quadratics, what's the next step? _____
       5. What steps find the final answer? _____
       _____

#2 1. What's the 1st step with equation fractions? $\dfrac{1}{x+1} + \dfrac{3}{5}$

              Solve it. What's the answer?   ___ + ___

       2. What's the 1st step with
          equation fractions? $\dfrac{x+5}{3x-9} + \dfrac{x+5}{5x-25}$

              Solve it. What's the answer?   ___ + ___

       3. What's the 1st step with
          equation fractions? $\dfrac{2}{x^2+8x+16} + \dfrac{3}{x^2+14x+49}$

              Solve it. What's the answer?   ___ + ___

#3 Solve for the variable.

1. $\dfrac{1}{a+3} + \dfrac{3}{5}$ $\dfrac{1}{b+2} + \dfrac{1}{4}$ Calculator? yes no

2. $\dfrac{x+6}{2x-4} + \dfrac{x-2}{4x-16}$ $\dfrac{x-5}{2x-8} + \dfrac{x-5}{6x-36}$

3. $\dfrac{4}{x^2+10x+25} + \dfrac{7}{x^2+18x+64}$ $\dfrac{1}{x^2+6x+9} + \dfrac{5}{x^2+18x+81}$

Review
1. What's the 1st step with an equation with fractions? _____
2. After writing the numbers over them, what's next? _____
3. What happens to what's left over? _____
4. If there are quadratics, what's the next step? _____
5. What steps find the final answer? _____

Calculator? yes no

Ch 9 Ls 2 How to change an equation's denominators. 73

_____ #1 #2 ____/ 9 #3 ____/ 8 R ____/ 10 T____/ 27 _____
  Name   Checker

Example. $\dfrac{x + \frac{x}{2}}{x + \frac{x}{3}}$

#1 1. What is the 1st step to solve this? _____

2. What is the 2nd step? _____

3. How do you solve 3x/2 over 4x/3? _____

4. The answer to the fraction? _____

5. How do you get same denominators? $\dfrac{x + \frac{x}{4}}{x + \frac{x}{5}}$

Same denominators. Combine them. $\dfrac{+\frac{x}{4}}{+\frac{x}{5}} = \dfrac{}{}$
Next step?

Divide. What is the final answer?   ── x ──
                                          ────

#2 1. What is a complex fraction? _____

2. How do you solve a complex fraction? _____

3. What rules does dividing fractions follow? _____

4. What's the 1st step to divide a complex fraction? $\dfrac{\frac{1}{x + 1}}{\frac{3}{2x + 2}}$

Finish it. What's the answer?   ─────── ÷ ───────

─────── • ─────── = ───────

#3 1. $\dfrac{\dfrac{t+3x}{w}}{6t} = \underline{\quad} \div \underline{\quad} \qquad \underline{\quad} \bullet \underline{\quad} = \underline{\quad}$ Calculator? yes   no

2. $\dfrac{\dfrac{10}{x}}{\dfrac{x}{2-s}} = \underline{\quad} \div \underline{\quad} \qquad \underline{\quad} \bullet \underline{\quad} = \underline{\quad}$

3. $\dfrac{\dfrac{1}{x}+4}{\dfrac{3}{x}+5} = \underline{\quad} \div \underline{\quad} \qquad \underline{\quad} \bullet \underline{\quad} = \underline{\quad}$

4. $\dfrac{\dfrac{x^2-9}{3}}{\dfrac{3-x}{5}} = \underline{\quad} \div \underline{\quad} \qquad \underline{\quad} \bullet \underline{\quad} = \underline{\quad}$

Example. $\dfrac{x+\dfrac{x}{2}}{x+\dfrac{x}{3}}$

Review 1. What is the 1st step to solve this? _____   Calculator?
2. What is the 2nd step? _____   yes   no
3. How do you solve 3x/2 over 4x/3? _____
4. The answer to the fraction? _____
5. What is a complex fraction? _____
6. How do you solve a complex fraction? _____
7. What rules does dividing fractions follow? _____

What's the answer? 8. $\dfrac{\dfrac{4x^2-16}{2x+2}}{\dfrac{8x-16}{4x}} = \underline{\quad} \div \underline{\quad} \qquad \underline{\quad} \bullet \underline{\quad} = \underline{\quad}$

9. $\dfrac{\dfrac{c+d}{4}}{\dfrac{c^2+d^2}{12}} = \underline{\quad} \div \underline{\quad} \qquad \underline{\quad} \bullet \underline{\quad} = \underline{\quad}$

10. $\dfrac{\dfrac{c+d}{4}}{\dfrac{c^2+d^2}{12}} = \underline{\quad} \div \underline{\quad} \qquad \underline{\quad} \bullet \underline{\quad} = \underline{\quad}$

Ch 9 Ls 3 How to change the denominators of an equation. 75

_____   #1 #2 ____ / 10 #3 ____ / 4 R ___ / 9 T ___ / 23 _____
              Name   Checker

#1   1. Name the 1st step to solve an entire equation of fractions. _____
_____

   2. What's the next step? _____
   3. What's left over for each term? _____
   4. How do you add fractions with binomials as denominators? _____
_____

   5. How about the denominator? _____
   6.   What's the 1st step with
        equation fractions? $\dfrac{1}{x+1} + \dfrac{2}{5} = \dfrac{x}{6}$

        Solve the next step. What's left over? $\dfrac{1}{x+1} + \dfrac{2}{5} = \dfrac{x}{6}$

        Solve it. What's the 1st step?   _____

        Take as many steps as you need. _____

                                         _____

                                         _____

#2   1. What does LCD stand for? _____
     2. What does it mean? _____
     3. What's the LCD? $\dfrac{1}{x+3} + \dfrac{3}{15} = \dfrac{x}{30(x+6)}$

     4. What's the LCD? $\dfrac{x+1}{x-1} + \dfrac{3x}{5} = \dfrac{1}{(x-1)}$

#3 1. $\dfrac{1}{a+2} + \dfrac{2}{6} = \dfrac{a}{6}$ $\dfrac{1}{b+1} + \dfrac{3}{4} = \dfrac{b}{7}$ Calculator? yes no

You need to solve for credit.

_____   _____

_____   _____

_____   _____

_____   _____

2. $\dfrac{1}{c+4} + \dfrac{3}{7} = \dfrac{c}{8}$ $\dfrac{1}{x+3} + \dfrac{1}{6} = \dfrac{x}{7}$

_____   _____

_____   _____

_____   _____

_____   _____

Review 1. Name the 1st step to solve an entire equation of fractions. _____   Calculator?
_____   yes no

2. What's the next step? _____

3. What's left over for each term? _____

4. How do you add fractions with binomials as denominators? _____

_____

5. How about the denominator? _____

6. What does LCD stand for? _____

7. What does it mean? _____

8. $\dfrac{1}{x^2+2x} + \dfrac{1}{x+2} = \dfrac{2}{x}$ LCD is _____

Find the LCD.

9. $\dfrac{4}{x^2+3x} + \dfrac{1}{x+3} = \dfrac{7}{x}$ LCD is _____

Review Problems   77

_____ #1 - #3 ____/ 9 Back ____/ 7 Total ____/ 16
           Name

1. Rational Equation _____

2. Partial Fraction _____

#2 1. $\dfrac{x-2}{3x-6} + \dfrac{x-2}{2x-4}$ $\dfrac{3}{4x-16} + \dfrac{4}{5x-20}$

_____   _____

_____   _____

_____   _____

_____   _____

3. $\dfrac{4}{x^2-2x-35} + \dfrac{7}{(x+5)}$ $\dfrac{2}{x^2-3x-40} - \dfrac{6}{(x-8)}$

_____   _____

_____   _____

_____   _____

_____   _____

#3 1. $\dfrac{x + \frac{x}{6}}{x + \frac{x}{7}} = \dfrac{+\frac{x}{6}}{+\frac{x}{7}} = \underline{\quad}$ $\underline{\quad} \bullet \underline{\quad} = \underline{\quad}$

Solve with
partial fractions.

2. $\dfrac{x + \frac{x}{8}}{x + \frac{x}{4}} = \dfrac{+\frac{x}{8}}{+\frac{x}{4}} = \underline{\quad}$ $\underline{\quad} \bullet \underline{\quad} = \underline{\quad}$

3. $\dfrac{x + \frac{x}{9}}{x + \frac{x}{7}} = \dfrac{+\frac{x}{9}}{+\frac{x}{7}} = \underline{\quad}$ $\underline{\quad} \bullet \underline{\quad} = \underline{\quad}$

#4

Solve the equation.

1. $\dfrac{1}{a+3} + \dfrac{3}{5} = \dfrac{4a}{10}$ $\dfrac{1}{b+2} + \dfrac{1}{4} = \dfrac{9}{20}$ Calculator? yes no

2. $\dfrac{1}{c+3} + \dfrac{3}{5} = \dfrac{2c}{5}$ $\dfrac{1}{x+2} + \dfrac{2}{5} = \dfrac{x}{16}$

#5

Simplify.

1. $\dfrac{\dfrac{6z^2 - 6}{8z^2 + 8}}{\dfrac{2z - 2}{4z^2 + 4z}}$ = ——— ÷ ———   ——— • ——— = ———   Calculator? yes no

2. $\dfrac{\dfrac{a + b}{3a - b}}{\dfrac{a + b}{3a + b}}$ = ——— ÷ ———   ——— • ——— = ———

3. $\dfrac{\dfrac{a + b}{3a - b}}{\dfrac{a + b}{3a + b}}$ = ——— ÷ ———   ——— • ——— = ———

Ch 10 Ls 1 Factor roots and match them. 79

_____ #1 #2 ____/ 11 #3 ____/ 12 R ____/ 14 T____/ 37 _____
          Name   Checker

#1 1. Name 3 steps to factor a square root. _____
_____

2. What happens with an odd exponent in a square root? _____

3. Match exponents. What factors out? $\sqrt{4x^2 y^2}$

_____

4. Match exponents. What factors out? $\sqrt[3]{9x^3 y^2}$

_____

5. Match exponents. What factors out? $\sqrt[4]{4x^2 y^4}$

_____

6. Match exponents. What factors out? $\sqrt[3]{8x^3 y^2}$

_____

#2 1. Why can a square root be both positive and negative? _____
2. Why does a square root make an absolute value? _____
3. Do even or odd roots make both positive negative? _____

4. Can you factor a negative? $\sqrt{-5x^2 y^2}$

_____

5. Can you factor a negative? $\sqrt{-8x^2 y^2}$

_____

#3 Solve the roots.

1. $\sqrt{196}$ $\sqrt{225}$ $\sqrt{0.81}$ Calculator? yes no

2. $\sqrt{4x^2y^2}$ $\sqrt{8x^3y^4}$ $\sqrt{(6g)^2}$

3. $\sqrt{49a^6}$ $\sqrt{x^2+2x+1}$ $\sqrt{81h^6}$

4. $\sqrt{x^2+6x+9}$ $\sqrt{20x^2y^2}$ $\sqrt{27x^3y^2}$

Review

1. Name 3 steps to factor a square root. _____   Calculator? yes no

2. What happens with an odd exponent in a square root? _____
3. Why can a square root be both positive and negative? _____
4. Why does a square root make an absolute value? _____
5. Do even or odd roots make both positive negative? _____

Solve polynomial roots.

6. $\sqrt[3]{27x^3y^3}$ $\sqrt[4]{x^4y^6z}$ $\sqrt[3]{8x^6y^3}$

7. $\sqrt[3]{(3x-1)^6}$ $\sqrt[3]{(5x+1)^3}$ $\sqrt{24x^2y^2}$

8. $\sqrt{12x^2y^6}$ $\sqrt[3]{(8x^3)^2}$ $\sqrt[4]{10,000}$

Ch 10 Ls 2 Solve binomial roots. 81

_____ #1 #2 ____/ 10 #3 ____/ 13 R ____/ 6 T ____/ 29 _____
         Name   Checker

#1 1. How does a Binomial square root solve? $\sqrt{x-3}^{\,2}$ _____
    2. How does square of 2 roots solve? $(\sqrt{2}-\sqrt{x})^2$ _____
    3. What do you use to multiply 2 squares of 2 roots? _____
    4. Write these roots so you can multiply them. $(\sqrt{x-2})^2$

                  What are the 1st 2 steps?   _____

                  What are the next 2 steps?   _____

                  What's the last, last step?   _____

                                                 _____

#2 1. What's the 1st step to get same roots? $4\sqrt{20}-2\sqrt{45}$ _____

    _____

    2. What's the 2nd step? _____
    3. Finish it. _____
    4. What rule does it use? _____
    5. What is the 1st step? $4\sqrt{28}-2\sqrt{63}$

                                                                    What's
            Change 28 to _____ and 63 to _____   2nd?

                             _____

    6. What is the 1st step? $6\sqrt{50}+2\sqrt{98}$

            Change 50 to _____ and 98 to _____   Finish it.

                             _____

#3 1. $2\sqrt{2} \cdot 3\sqrt{8}$ $3\sqrt{3} \cdot 9\sqrt{12}$ $4\sqrt{3} \cdot 8\sqrt{7}$ Calculator? yes no

Multiply these mixed roots. Simplify.

_____   _____   _____

_____   _____   _____

2. $2\sqrt{x} \cdot 3\sqrt{x}$ $4\sqrt{x} \cdot 6\sqrt{x^2}$ $5\sqrt{x} \cdot 3\sqrt{x^2}$

_____   _____   _____

_____   _____   _____

3. $\sqrt{8x}(\sqrt{2} + 2\sqrt{x})$ $\sqrt{8x}(\sqrt{2} + 2\sqrt{x})$ $\sqrt{8x}(\sqrt{2} + 2\sqrt{x})$

_____   _____   _____

_____   _____   _____

4. $(\sqrt{x^3} \cdot \sqrt{x^5})$ $(\sqrt[3]{x^2} \cdot \sqrt{x^5})$

_____   _____

_____   _____

5. $2\sqrt{48} - 5\sqrt{45}$ $4\sqrt{20} + 2\sqrt{45}$

_____   _____

_____   _____

Review 1. How does a Binomial square root solve? $\sqrt{x-3}^2$ _____ Calculator? yes no

2. How does square of 2 roots solve? $(\sqrt{2} \cdot \sqrt{x})^2$ _____

3. What do you use to multiply 2 squares of 2 roots? _____

4. What's the 1st step to get same roots? $4\sqrt{20} - 2\sqrt{45}$ _____

_____

5. Finish it. _____

6. What rule does it use? _____

Ch 10 Ls 3 Decide if it uses the Ma or Me rule. 83

_____ #1 #2 ____ / 10 #3 ____ / 18 R ____ / 4 T ____ / 32 _____
        Name   Checker

#1 1. How do you solve multiplication with same bases? _____
2. What rule solves multiplication with same exponents? _____
3. What is the DS Rule? _____
4. How do you solve a fraction with same bases and root exponents? _____
_____

#2 1. Which rule is it? Solve it. $3^{2\sqrt{3}} \cdot 3^{\sqrt{3}}$

Ma or Me Rule   _____

2. Which rule is it? Solve it. $5^{\frac{1}{2}} \cdot 2^{\frac{1}{2}}$

Ma or Me Rule   _____

3. Which rule is it? Solve it. $6^{\frac{1}{2}} \cdot 5^{\frac{1}{2}}$

Ma or Me Rule   _____

4. Which rule is it? Solve it. $3^{2\sqrt{3}} \cdot 3^{\sqrt{3}}$

Ma or Me Rule   _____

5. Which rule is it? Solve it. $3^{\frac{1}{3}} \cdot 8^{\frac{1}{3}}$

Ma or Me Rule   _____

6. Which rule is it? Solve it. $5^{2\sqrt{6}} \cdot 5^{\sqrt{6}}$

Ma or Me Rule   _____

#3 Divide the fraction roots.   Calculator?
                                                                     yes   no

1. $6\sqrt{15} \div 2\sqrt{5}$ $8\sqrt{25} \div 2\sqrt{5}$ $9\sqrt{30} \div 3\sqrt{6}$

   _____   _____   _____

2. $\dfrac{2\sqrt{8}}{3\sqrt{2}}$ $\dfrac{2\sqrt{5}}{6\sqrt{15}}$ $\dfrac{9\sqrt{32}}{6\sqrt{8}}$

   _____   _____   _____

3. $\sqrt{x^2 \div x^4}$ $\sqrt{\dfrac{4x}{2x^2}}$ $\sqrt{\dfrac{8x^2}{2x}}$

   _____   _____   _____

4. $\sqrt{a^5 \div a^4}$ $\sqrt{c^3 \div c^2}$ $\sqrt{x^2 \div x^4}$

Solve with
the DS Rule.   _____   _____   _____

5. $\sqrt{\dfrac{6x}{x^2}}$ $\sqrt{\dfrac{18a^2}{2a^2}}$ $\sqrt{\dfrac{4b^2}{16b^3}}$

   _____   _____   _____

6. $\dfrac{2\sqrt{8}}{3\sqrt{2}}$ $\dfrac{2\sqrt{8}}{3\sqrt{2}}$ $\dfrac{2\sqrt{8}}{3\sqrt{2}}$

   _____   _____   _____

Review 1. What happens when there's a root in a denominator? _____   Calculator?
        2. How do you rationalize the denominator? _____   yes   no
        3. How do you solve a denominator with a binomial root? _____
        4. What happens to the middle terms? _____

Ch 10 Ls 4 Ds, Simplify Fractions and Ratioanlize Denominators 85

_____ #1 #2 ____/ 10 #3 ____/ 10 R ____/ 11 T ____/ 31 _____
        Name   Checker

#1 1. Root divided by root equals what kind of answer? _____

     2. What root rule uses variables? _____

     3. How do mixed roots divide? _____

     4. How do you divide SR 2a^3/ SR 8a? _____

     5. What's the 1st step to solve this fraction? $\dfrac{15\sqrt{a^4}}{5\sqrt{a^5}}$

Simplify the numbers. What's the 2nd step? $\dfrac{\sqrt{}}{\sqrt{}}$

Ds same variables. $\dfrac{\overline{}}{\sqrt{}}$

#2 1. What happens when there's a root in a denominator? _____

     2. How do you rationalize the denominator? _____

     3. How do you solve a denominator with a binomial root? _____

     4. What happens to the middle terms? _____

     5. What's the 1st step to solve a root in a denominator? $\dfrac{\sqrt{2}}{\sqrt{x}-1}$

What's the 1st step to multiply it? $\dfrac{\sqrt{2}}{\sqrt{x}-1} \cdot \dfrac{}{}$

What's the 2nd step?   _____

Finish it. What's the answer?   _____

                                                                             _____

1. $\dfrac{35\sqrt{a^5}}{20\sqrt{a^4}} = \underline{}$ $\dfrac{45\sqrt{a^6}}{5\sqrt{a^2}} = \underline{}$ Calculator? yes no

Simplify it

2. $\dfrac{\sqrt{6}}{2\sqrt{3}} = \underline{}$ $\dfrac{\sqrt{8}}{7\sqrt{2}} = \underline{}$

3. $\dfrac{\sqrt{z+1}}{\sqrt{z-1^2}} = \underline{}$ $\dfrac{72\sqrt{x}}{8\sqrt{x^7}} = \underline{}$

4. $\dfrac{65\sqrt{b}}{15\sqrt{b^5}} = \underline{}$ $\dfrac{28\sqrt{c^3}}{12\sqrt{c}} = \underline{}$

5. $\dfrac{18\sqrt{9}}{12\sqrt{3}} = \underline{}$ $\dfrac{6\sqrt{40}}{4\sqrt{35}} = \underline{}$

Review 1. Root divided by root equals what kind of answer? _____ Calculator? yes no

2. What root rule uses variables? _____

3. How do mixed roots divide? _____

4. How do you divide SR 2a^3/ SR 8a? _____

5. What happens when there's a root in a denominator? _____

6. How do you rationalize the denominator? _____

7. How do you solve a denominator with a binomial root? _____

8. What happens to the middle terms? _____

9. Solve these roots. $\dfrac{\sqrt{5}}{\sqrt{x-2}}$ $\dfrac{1}{\sqrt{x+1}}$ $\dfrac{1}{\sqrt{8}}$

____ • ____   ____ • ____   ____ • ____

_____   _____   _____

_____   _____   _____

Review Problems

_____ #1 - #3 ____ / 19   Back ____ / 21   Total ____ / 40
Name

1. Match Exponents _____
2. Nth Root _____
3. Rationalize the Denominator _____

#2
Simplify.

1. $\sqrt[3]{8x^3 y}$ = _____ $\sqrt[4]{16x^8 y^4}$ = _____

2. $\sqrt[3]{16x^6 y^2}$ = _____ $\sqrt[4]{81x^2 y^2}$ = _____

3. $\sqrt[3]{54x^4 y^3}$ = _____ $\sqrt[3]{27x^6 y^6}$ = _____

#3
Multiply these mixed roots.

1. $5\sqrt{x} \cdot 7\sqrt{x}$ $6\sqrt{x} \cdot 8\sqrt{x^3}$ $9\sqrt{x} \cdot 3\sqrt{x^2}$

   _____   _____   _____

2. $\sqrt{8x}(\sqrt{2} + 2\sqrt{x})$ $\sqrt{8x}(\sqrt{2} + 2\sqrt{x})$ $\sqrt{8x}(\sqrt{2} + 2\sqrt{x})$

   _____   _____   _____
   _____   _____   _____

3. $(\sqrt{3x} + \sqrt{2x})(\sqrt{4x} + \sqrt{5x})$ $(\sqrt{x^2} + \sqrt{x^5})(\sqrt{x^2} - \sqrt{x^5})$

   _____   _____
   _____   _____

4. $(\sqrt{5a} + \sqrt{4a})(\sqrt{2a} - \sqrt{8a})$ $(\sqrt{9b} - \sqrt{3b})(\sqrt{5b} - \sqrt{7b})$

   _____   _____
   _____   _____

#3 1. $9\sqrt{35} \div 3\sqrt{5} =$ _____ $6\sqrt{16} \div 2\sqrt{8} =$ _____   Calculator? yes no

2. $4\sqrt{42} \div 2\sqrt{7} =$ _____ $6\sqrt{48} \div 3\sqrt{3} =$ _____

Divide the fraction roots.

3. $x^2 \div x^4 =$ _____ $\dfrac{4x}{2x^2} =$ _____

4. $\dfrac{10x}{2x^2} =$ _____ $a^7 \div a^4 =$ _____

5. $\dfrac{8b^2}{2b^3} =$ _____ $\dfrac{2\sqrt{8}}{3\sqrt{2}} =$ _____

6. $\dfrac{2\sqrt{8}}{6\sqrt{2}} =$ _____ $\dfrac{8a^3}{2a^2} =$ _____

#3 What does it multiply to solve a root in a denominator?   Calculator? yes no

1. $\dfrac{1}{\sqrt{5}}$ x _____ $\dfrac{1}{\sqrt{7}}$ x _____ $\dfrac{1}{\sqrt{6}}$ x _____

What does it equal?   _____   _____   _____

2. $12^{-\frac{1}{2}}$ $27^{-\frac{1}{2}}$ $64^{-\frac{1}{2}}$

Solve the fraction exponent.   _____   _____   _____

Solve the negative exponent.   _____   _____   _____

3. $2^{\frac{3}{2}}$ $3^{\frac{4}{3}}$ $b^{\frac{3}{4}}$

Write the problem as 2 numbers.   _____   _____   _____

Solve both numbers.   _____   _____   _____

What does it equal?   _____   _____   _____

Ch 11 Ls 1 Change the Base 2 Ways   89

_____ #1 #2 ____ / 8 #3 ____ / 9 R ____ / 6 T ____ / 23 _____
        Name   Checker

#1 1. How can different bases be changed? $\dfrac{8^{\frac{1}{6}}}{2^{\frac{1}{3}}}$ _____

    2. How do different fractions exponents solve? $2^{\frac{3}{6} - \frac{1}{3}}$ _____

    3. How does it get same bases? $\dfrac{8^{\frac{5}{6}}}{2^{\frac{1}{3}}}$

    _____

    _____

    4. How do you solve
       different fraction exponents? $5^{\frac{3}{4} - \frac{1}{2}}$

    _____

#2 1. What if the denominator can't make same bases? _____
    2. What fraction does it multiply? _____
    3. What does this
       fraction multiply? $\dfrac{\sqrt[3]{2}}{\sqrt[3]{7x^2}}$ $\dfrac{\sqrt[3]{8}}{\sqrt[3]{x+1}}$

       _____   _____

       _____   _____

       _____   _____

#3 1. $\dfrac{64^{\frac{1}{6}}}{4^{\frac{1}{3}}}$ $\dfrac{27^{\frac{1}{6}}}{3^{\frac{1}{3}}}$ $\dfrac{125^{\frac{3}{4}}}{5^{\frac{1}{2}}}$ Calculator? yes no

How does it get same bases? _____   _____   _____

_____   _____   _____

2. $9^{\frac{5}{6}-\frac{1}{2}}$ $6^{\frac{3}{4}-\frac{1}{2}}$ $2^{\frac{7}{8}-\frac{1}{4}}$

Solve different fraction exponents? _____   _____   _____

_____   _____   _____

3. $\dfrac{\sqrt[3]{7}}{\sqrt[3]{4x^2}}$ $\dfrac{\sqrt[3]{2}}{\sqrt[3]{3x^2}}$ $\dfrac{\sqrt[3]{4}}{\sqrt[3]{5x^2}}$

What does this fraction multiply? _____   _____   _____

_____   _____   _____

Review 1. What if the denominator can't make same bases? _____   Calculator? yes no

2. What fraction does it multiply? _____

3. How can different bases be changed? $\dfrac{8^{\frac{1}{6}}}{2^{\frac{1}{3}}}$ _____

4. How do different fractions exponents solve? $2^{\frac{3}{6}-\frac{1}{3}}$ _____

_____

5. What if the denominator can't make same bases? $\dfrac{\sqrt[3]{5}}{\sqrt[3]{2x^2}}$ _____

_____

6. What fraction does it multiply? _____

Ch 11 Ls 2   Solve equations with roots.   91

_____ #1 #2 ____ / 5   #3 ____ / 4   R ____ / 3   T ____ / 12 _____
           Name   Checker

#1 1. What's the 1st step to solve different exponents on 1 side? _____
_____

2. How do you solve a square root on 1 side? _____
3. What are 3 steps to find X intercepts? _____
_____

#2 1. What's the 1st step with different exponents? $x + 3 = x\sqrt{5}$

What happens with the sides?   _____

Solve the squares.   _____

What's the next step?   _____

How does it get a final answer?   _____

What are the answers?   _____

                                   _____

2. What's the 1st step with different exponents? $\sqrt{3x + 15} - 5 = x$

What happens with the sides?   _____

Solve the squares.   _____

What's the next step?   _____

How does it get a final answer?   _____

What are the answers?   _____

                                   _____

#3 1. $\sqrt{0.4x + 20} - 2 = x$ $x + 4 = x\sqrt{8}$ Calculator? yes no

_____   _____
_____   _____
_____   _____
_____   _____
_____   _____
_____   _____

2. $x + 6 = x\sqrt{3}$ $\sqrt{6x + 8} - 4 = x$

_____   _____
_____   _____
_____   _____
_____   _____
_____   _____
_____   _____

Review 1. What's the 1st step to solve different exponents on 1 side? _____   Calculator? yes no
_____

2. How do you solve a square root on 1 side? _____

3. What are 3 steps to find X intercepts? _____
_____

Review Problems   93

_____   #1 - #3 ____ / 14   Back ____ / 6   Total ____ / 20
Name

1. Change the Base 2 Ways _____
2. Rationalize Square Root _____

#2 Solve the bases with different exponents.

1.

$$\dfrac{343^{\frac{1}{6}}}{7^{\frac{1}{3}}} \qquad \dfrac{729^{\frac{1}{6}}}{9^{\frac{1}{3}}} \qquad \dfrac{512^{\frac{1}{6}}}{8^{\frac{1}{3}}}$$

How does it get same bases?

_____   _____   _____

_____   _____   _____

2. $\quad 5^{\frac{5}{9} - \frac{1}{3}} \qquad 8^{\frac{7}{8} - \frac{1}{2}} \qquad 7^{\frac{5}{6} - \frac{1}{2}}$

_____   _____   _____

_____   _____   _____

3. $\quad \dfrac{\sqrt[3]{7}}{\sqrt[3]{x+1^{2}}} \qquad \dfrac{\sqrt[3]{4}}{\sqrt[3]{6x^{2}}} \qquad \dfrac{\sqrt[4]{3}}{\sqrt[4]{5x^{2}}}$

_____   _____   _____

_____   _____   _____

4. $\quad \dfrac{\sqrt[3]{2}}{\sqrt[3]{7x^{2}}} \qquad \dfrac{\sqrt[3]{6}}{\sqrt[3]{x+3^{2}}} \qquad \dfrac{\sqrt[4]{5}}{\sqrt[4]{2x^{2}}}$

_____   _____   _____

_____   _____   _____

#3
Solve these.

Calculator? yes   no

1. $\sqrt{8x+12} - 6 = x$

   $x + 6 = x\sqrt{7}$

2. $\sqrt{5x+14} - 2 = x$

   $x + 2 = x\sqrt{6}$

3. $\sqrt{7x+20} - 5 = x$

   $x + 3 = x\sqrt{8}$

Ch 12 Ls 1 Begin imaginary numbers. 95

_____ #1 #2 ____ / 10 #3 #4 ____ / 18 R ___ / 13 T ___ / 41 _____
         Name   Checker

#1 1. Why do we need imaginary numbers? _____

2. What symbol takes it's place? _____

3. Name 2 steps to factor $\sqrt{-12}$. _____

4. What's 1st step to factor square root of - 8? $\sqrt{-9}$

How does it use an imaginary number?   _____

5. What's 1st to factor negative perfect roots? $\sqrt{-3}$

How does it use an imaginary number?   _____

6. What is 5 i in root form?   **5 i**

                                                _____

#2 1. How does square root of negative X cubed simplify? $\sqrt{-x^3}$ _____

2. How do you solve a negative variable in a square root? _____

3. What's the 1st step to solve this. $\sqrt{-12x^2}$

Finish it. What's the answer?   _____

4. What's the 1st step to solve this. $\sqrt{-12x^2}$

Finish it. What's the answer?   _____

#3 **Find imaginary numbers.**

1. $\sqrt{-12} =$ _____ $\sqrt{-5} =$ _____ $\sqrt{-8} =$ _____   Calculator? yes no

   _____   _____   _____

2. $\sqrt{-10} =$ _____ $\sqrt{-20} =$ _____ $\sqrt{-25} =$ _____

   _____   _____   _____

3. $\sqrt{-6} =$ _____ $\sqrt{-40} =$ _____ $\sqrt{-45} =$ _____

   _____   _____   _____

#4

1. $\sqrt{-x^3} =$ _____ $\sqrt{-x^4} =$ _____ $\sqrt{-x^5} =$ _____

2. $\sqrt{-3x^2} =$ _____ $\sqrt{-10x^2} =$ _____ $\sqrt{-8x^2} =$ _____

   _____   _____   _____

3. $\sqrt{-16x^2} =$ _____ $\sqrt{-15x^2} =$ _____ $\sqrt{-20x^2} =$ _____

   _____   _____   _____

Review

1. Why do we need imaginary numbers? _____   Calculator? yes no
2. How do you write an imaginary number? _____
3. How does square root of negative X cubed simplify? $\sqrt{-x^3}$ _____
4. How do you solve a negative variable in a square root? _____

Change to negative roots.

5. $2i\sqrt{3} =$ _____ $4i\sqrt{3} =$ _____ $3i\sqrt{3} =$ _____

6. $4i\sqrt{2} =$ _____ $2i\sqrt{2} =$ _____ $5i\sqrt{2} =$ _____

7. $2i\sqrt{5} =$ _____ $3i\sqrt{4} =$ _____ $3i\sqrt{6} =$ _____

Ch 12 Ls 2 Imaginary numbers as expressions. 97

_____ #1 #2 ____ / 11 #3 #5 ____ / 18 R ____ / 17 T ____ / 46 _____
           Name   Checker

#1 1. How do you write an imaginary number with a number? _____

2. How do you add imaginary numbers? $(3 + 2i) + (1 + 5i)$ _____

_____

3. How is subtracting them different? _____

    4. Make a simplified root answer. $\sqrt{27}$

                                                                        _____

    5. How does it use an imaginary number? $4 + 8i$

                                                                          _____

    6. Add the binomilals. $(3 + 2i) + (1 + 5i)$

                                                                          _____

    7. Subtract the binomilals. $(4 + 3i) - (2 + 4i)$

                                                                           _____

#2 1. How does square root of a negative X cubed simplify? $\sqrt{-x^3}$ _____

2. How do you solve a negative variable in a square root? _____

3. What's 1st step to factor square root of - 8? $\sqrt{-8}$

How does it use an imaginary number?   _____

                                                                          _____

4. What's 1st to factor negative perfect roots? $\sqrt{-9x^2}$

How does it use an imaginary number?   _____

                                                                          _____

#3 Write these as simplified binomials.   Calculator? yes no

1. $2\sqrt{-5} = $ _____ $6\sqrt{-10} = $ _____ $3\sqrt{-8} = $ _____

2. $7\sqrt{-6} = $ _____ $5\sqrt{-3} = $ _____ $2\sqrt{12} = $ _____

#4 Change to negative roots.

1. $2 + 3i = $ _____ $1 + 5i = $ _____ $4 + 6i = $ _____

2. $1 + 2i = $ _____ $3 + 4i = $ _____ $2 + 4i = $ _____

3. $8 + 3i = $ _____ $7 + 3i = $ _____ $4 + 5i = $ _____

#5 Add or subtract the binomials.

1. $(2 + 3i) + (3 + 5i) = $ _____ $(4 - 2i) + (6 + 4i) = $ _____

2. $(6 - 4i) + (2 - 3i) = $ _____ $(6 + 3i) - (2 + 2i) = $ _____

3. $(3 + 7i) - (3 - 5i) = $ _____ $(4 + 3i) - (1 + 5i) = $ _____

Review 1. How do you write an imaginary number with a number? _____   Calculator? yes no

2. How do you add imaginary numbers? $(3 + 2i) + (1 + 5i)$ _____

_____

3. How is subtracting them different? _____

4. How does square root of negative X cubed simplify? $\sqrt{-x^3}$ _____

5. How do you solve a negative variable in a square root? _____

6. $2xi\sqrt{3} = $ _____ $4xi\sqrt{3} = $ _____ $3xi\sqrt{3} = $ _____

7. $2i\sqrt{3x} = $ _____ $4i\sqrt{3x} = $ _____ $3i\sqrt{3x} = $ _____

8. $\sqrt{-3x^2} = $ _____ $\sqrt{-10x^2} = $ _____ $\sqrt{-8x^2} = $ _____

9. $\sqrt{-14x^2} = $ _____ $\sqrt{-15x^2} = $ _____ $\sqrt{-20x^2} = $ _____

Ch 12 Ls 3   Multiply imaginary numbers.   99

_____ #1 #2 ____ / 7   #3 #4 ____ / 15   R ____ / 3   T ____ / 25   _____
Name   Checker

#1   1. How do you multiply binomials with imaginary numbers? _____
      2. How does i squared simplify? _____
      3. If a trinomial has i squared, what does it equal? _____

#2   1. Multiply these imaginary numbers.   **2 i · 9 i**

_____

      2. Multiply these.   **- 3 i · 8 i**

_____

      3. Multiply the 1st 2 steps of these binomials.   **(3 + 2i)(1 + 5i)**

      Find the next 2 steps.   _____ _____

      What is the final answer?   _____ _____

_____

_____

      4. Multiply the 1st 2 steps of these binomials.   **(2 - 2i)(3 + 6i)**

      Find the next 2 steps.   _____ _____

      What is the final answer?   _____ _____

_____

_____

#3 How to multiply imaginary numbers. Calculator? yes no

1. $-2i \cdot 5i =$ _____ $3i \cdot 9i =$ _____ $4i \cdot 8i =$ _____

2. $5i \cdot 5i =$ _____ $-2i \cdot 7i =$ _____ $3i \cdot 6i =$ _____

3. $4i \cdot 3i =$ _____ $6i \cdot 8i =$ _____ $3i \cdot 9i =$ _____

#4 Use foil to Factor.

1. $(4 + 2i)(1 + 4i)$ $(7 - 2i)(1 + 5i)$ Calculator? yes no

2. $(3 + 2i)(2 - 3i)$ $(6 - 2i)(1 - 2i)$

Simplify and multiply.

3. $2\sqrt{-8} \cdot 3\sqrt{-12}$ $3\sqrt{-6} \cdot 2\sqrt{-6}$

Review
1. How do you multiply binomials with imaginary numbers? _____   Calculator? yes no
2. How does i squared simplify? _____
3. If a trinomial has i squared, what does it equal? _____

_____ #1 - #5 ____ / 26   Back ____ / 12   Total ____ / 38
Name

1. Complex Number _____

2. Conjugate _____

#2 What are imaginary numbers?

1. $\sqrt{-12}$ = _____ $\sqrt{-5}$ = _____ $\sqrt{-8}$ = _____

2. $\sqrt{-10}$ = _____ $\sqrt{-20}$ = _____ $\sqrt{-25}$ = _____

#3 Write these as simplified binomials.

1. $2\sqrt{5}$ = _____ $6\sqrt{-10}$ = _____ $3\sqrt{-8}$ = _____

2. $7\sqrt{-6}$ = _____ $5\sqrt{-3}$ = _____ $2\sqrt{12}$ = _____

#4 Change to negative roots.

1. $2 + 3i$ = _____ $1 + 5i$ = _____ $4 + 6i$ = _____

2. $1 + 2i$ = _____ $3 + 4i$ = _____ $2 + 4i$ = _____

3. $8 + 3i$ = _____ $7 + 3i$ = _____ $4 + 5i$ = _____

#5 Multiply these.

1. $5i \cdot 5i$ = _____ $-2i \cdot 7i$ = _____ $3i \cdot 6i$ = _____
   _____   _____   _____

2. $4i \cdot 3i$ = _____ $6i \cdot 8i$ = _____ $3i \cdot 9i$ = _____
   _____   _____   _____

3. $(4 + 2i)(1 + 4i)$ $(7 - 2i)(1 + 5i)$

   ___ ___ ___ ___   ___ ___ ___ ___

   _____   _____

   _____   _____

#6 Factor a fraction with imaginary numbers.   Calculator?
                                                                                                                  yes   no

1. $\dfrac{6i}{4i}$ $\dfrac{12i}{2i}$ $\dfrac{13}{4i}$

    _____   _____   _____

    _____   _____   _____

2. $\dfrac{3i}{5+i}$ $\dfrac{8i}{2+i}$ $\dfrac{4i}{1+i}$

    _____   _____   _____

    _____   _____   _____

    _____   _____   _____

#7 Solve these equations.   Calculator?
                                                              yes   no

1. $9x^3 + 24 = 0$ $3x^2 + 17 = 0$

     _____   _____

     _____   _____

     _____   _____

2. $5x^2 + 21 = 0$ $4x^2 + 19 = 0$

     _____   _____

     _____   _____

     _____   _____

3. $6x^2 + 18 = 0$ $2x^2 + 28 = 0$

     _____   _____

     _____   _____

     _____   _____

Ch 13 Ls 1   Imaginary numbers with larger exponents.   103

_____ #1 #2 ____ / 8   #3 ____ / 9   R ____ / 7   T ____ / 24 _____
          Name   Checker

#1 1. How does i squared simplify? _____

2. How does i squared use i to the 6th? _____

3. What if the imaginary number has an odd exponent? _____

4. How do you factor i squared to the 4th? i^4

                                                   _____

5. How do you factor i squared to the 5th? i^5

                                                   _____

#2 1. How do you graph complex number points? _____

2. How does the graph add them? _____

3. How do you graph these points?   **- 2 - 3i**   **4 + i**

- 2 - 3i
4 + 1i
‾‾‾‾‾‾

How do you add the points?

- 2 - 3i
4 + 1i
‾‾‾‾‾‾

Add the binomials.

#3 Factor these.

Calculator?
yes   no

1. i^{10} i^8 i^9

    _____   _____   _____

2. i^7 i^{11} i^5

    _____   _____   _____

3. i^6 i^4 i^3

    _____   _____   _____

Review 1. How does i squared simplify? _____

2. How does i squared use i to the 6th? _____

3. What if the imaginary number has an odd exponent? _____

4. How do you graph complex number points? _____

5. How does the graph add them? _____

Calculator?
yes   no

6. How do you graph these points?
Add them on the graph.

$3 + i$
$\underline{-4 + 3i}$

7. How do you graph these points?
Add them on the graph.

$4 + 2i$
$\underline{-1 + i}$

Ch 13 Ls 2 Imaginary numbers in fractions. 105

_____ #1 #2 ____ / 7 #3 ____ / 9 R ____ / 4 T ____ / 20 _____
                Name   Checker

#1 1. Can you leave slanted i in a denominator? _____
        2. How do you solve i in a denominator? _____
        3. If a binomial has i squared, what does it mean? _____
        4. What do you use to solve a binomial in the denominator? _____

#2 1. What is the 1st step to factor a
           fraction with imaginary numbers? $\dfrac{2 + i}{4i}$

           What's the next step?   _____

           Finish the fractions.
                                                    _____

                                                    _____

        2. What is the 1st step to factor a fraction? $\dfrac{4i}{2 + i}$

           What's the next step?   _____

           Finish the fractions.   _____

                                                    _____

                                                    _____

        3. What is the 1st step to factor a fraction? $\dfrac{3 + i}{6i}$

           What's the next step?   _____

           Finish the fractions.   _____

                                                    _____

#3 Factor a fraction with imaginary numbers.   Calculator?
yes   no

1. $\dfrac{6i}{4i}$ $\dfrac{12i}{2i}$ $\dfrac{13}{4i}$

_____   _____   _____
_____   _____   _____
_____   _____   _____

2. $\dfrac{3i}{5+i}$ $\dfrac{8i}{2+i}$ $\dfrac{4i}{1+i}$

_____   _____   _____
_____   _____   _____
_____   _____   _____
_____   _____   _____

3. $\dfrac{5+i}{8i}$ $\dfrac{6+i}{3i}$ $\dfrac{3+i}{2i}$

_____   _____   _____
_____   _____   _____
_____   _____   _____

Review 1. Can you leave slanted i in a denominator? _____   Calculator?
2. How do you solve i in a denominator? _____   yes   no
3. If a binomial has i squared, what does it mean? _____
4. What do you use to solve a binomial in the denominator? _____

Ch 13 Ls 3 Equations and imaginary numbers. 107

_____ #1 #2 ____ / 7   #3 ____ / 6   R ___ / 4   T ____ / 17 _____
Name   Checker

#1 1. What equation makes a negative square root? _____

2. Why does $x^2 + 1 = 0$ make an imaginary number? _____

3. Why does $x^2 = -1$ make a + or - ? _____

4. Does the + or - stay with the imaginary number? _____

#2 1. Solve the 1st step of this equation. $2x^2 + 50 = 0$

What's the next step? _____

How does it use an imaginary number? _____

_____

2. Solve the 1st step of this equation. $x^3 + 27 = 0$

What's the next step? _____

How does it use an imaginary number? _____

_____

3. Solve the 1st step of this equation. $x^4 + 16 = 0$

What's the next step? _____

How does it use an imaginary number? _____

_____

#3 Solve these equations.   Calculator?
                                                                yes   no

1. $9x^3 + 24 = 0$ $3x^2 + 17 = 0$

    _____   _____

    _____   _____

    _____   _____

2. $5x^2 + 21 = 0$ $4x^2 + 19 = 0$

    _____   _____

    _____   _____

    _____   _____

3. $8x^2 + 24 = 0$ $7x^4 + 32 = 0$

    _____   _____

    _____   _____

    _____   _____

Review 1. What equation makes a negative square root? _____   Calculator?
        2. Why does this equation make an imaginary number? $x^2 + 1 = 0$ _____   yes   no
        _____
        3. Why does $X^2 = -1$ make a + or - ? _____
        4. Does the + or - stay with the imaginary number? _____

Ch 13 Ls 4 Imaginary number word problems. 109

_____ #1 #2 ____ / 11 #3 ____ / 3 R ____ / 7 T ____ / 21 _____
          Name   Checker

#1 1. How does electricity use complex numbers? _____
       2. What does O stand for in resistance numbers? _____
       3. How do you decide if a set of electric plugs meets code? _____
       _____

       4. You have 2 circuits with these ohm readings.   6 - 8 j
          What is the combined resistance?   5 + 11 j
                                                              _____
            Add them to make sure
            it's under the level for the code.   +   j

#2 1. What's the current formula? _____
       2. What does resistance do? _____
       3. What measures current, resistance, and volts? _____
       4. Which parts of the current formula use imaginary numbers? _____
       _____

       5. A 120 volt circuit has amperage of **15 - 3j**.   What is the formula?
          What is the impedance?

                    Amps Ohms Volts
                     i x R = V   Where do the numbers go?
                    Current Resistance Voltage

                  (15 - 3j) x R = 120   What's the 1st step?

                          120
                  R = ─────────────   What's the next step?
                       (- 3j)

                          120   (+ 3j)   What's the
                  R = ─────────── • ──────────   denominator equal?
                       (- 3j)   (+ 3j)

                 120   (+ 3j)   ?
          R = ─────── • ──────────── = ───────   What's the answer?
              (- 3j)   (+ 3j)

                 120   (+ 3j)
          R = ─────── • ──────────── = ─────── =
              (- 3j)   (+ 3j)

#3 1. You have 2 circuits with these ohm readings.   6 - 8 j   Calculator?
       What is the combined resistance?   5 + 11 j   yes   no

       Add them to make sure it's under the level for the code.   ___ + ___ j

2. A 110 volt circuit has amperage of **10 - 2j**.
   What is the impedance?

   (10 - 2j) x R = 110

3. A 220 volt circuit has amperage of **20 - 4j**.
   What is the impedance?

   (20 - 4j) x R = 220

3. A 120 volt circuit has amperage of **8 - 3j**.
   What is the impedance?

   (8 - 3j) x R = 120

Review
1. How does electricity use complex numbers? _____   Calculator?
2. What does O stand for in resistance numbers? _____   yes   no
3. How do you decide if a set of electric plugs meets code? _____
4. What's the current formula? _____
5. What does resistance do? _____
6. What measures current, resistance, and volts? _____
7. Which parts of the current formula use imaginary numbers? _____

Review Problems 111

_____ #1 - #4 ____ / 15 Back ____ / 9 Total ____ / 24
Name

1. Complex Number _____

2. Conjugate _____

#2 1. i^{10} i^{8} i^{9} i^{3}
Factor
these.
    _____   _____   _____   _____

  2. i^{7} i^{11} i^{5} i^{4}

    _____   _____   _____   _____

#3 1. How do you graph these points?
Graph   Add them on the graph.
these.
         $3 + i$
         $-4 + 3i$
         ‾‾‾‾‾‾‾

2. How do you graph these points?
   Add them on the graph.

         $4 + 2i$
         $-1 + i$
         ‾‾‾‾‾‾‾

#4 1. $\dfrac{6i}{4i}$ $\dfrac{12i}{2i}$ $\dfrac{13}{4i}$
Factor
these.

    _____   _____   _____
    _____   _____   _____
    _____   _____   _____

#4 2. Continued.

$$\frac{3i}{5+i} \qquad \frac{8i}{2+i} \qquad \frac{4i}{1+i}$$

Calculator? yes no

#5 Solve these equations.

1. $9x^3 + 24 = 0$ \qquad $3x^2 + 17 = 0$

Calculator? yes no

2. $5x^2 + 21 = 0$ \qquad $4x^2 + 19 = 0$

#6

1. A 110 volt circuit has amperage of **12 - 3j**. What is the impedance?

$$(12 - 3j) \times R = 110$$

Calculator? yes no

2. A 220 volt circuit has amperage of **30 - 8j**. What is the impedance?

$$(30 - 8j) \times R = 220$$

Ch 14 Ls 1 Polynomial Functions 113

_____ #1 #2 ____/ 12 #3 ____/ 12 R ____/ 9 T____/ 33 _____
          Name   Checker

#1 1. What is a function? _____

2. What is a vertical line test? _____

3. How are discrete functions different from continuous ones? _____

4. What is the domain and range of a function? _____

5. How does a function show the 4 operations? _____

#2 1. Is this a function?   1, 3   2, 3   3, 3

Yes, it is a function.   No, it is not a function.

2. Is this a function?   1, 3   1, 3   3, 3

Yes, it is a function.   No, it is not a function.

3. What is the domain? Find the range.

Domain: _____   Range: _____

4. Add $(a + b)(x)$

$a(x) = 2x + 2$
$b(x) = 5x - 5$

$a(x) = -4x + 2$
$b(x) = 7x - 1$

5. Subtract $(a - b)(x)$

6. Multiply $(a \bullet b)(x)$

7. Divide $(\frac{a}{b})(x)$

#3 Add, Subtract, Multiply, and Divide Functions   Calculator? yes no

$a(x) = 3x + 1$ $a(x) = -3x + 4$
$b(x) = 8x - 2$ $b(x) = 9x - 2$

1. Add $(a + b)(x)$ _____   _____

2. Subtract $(a - b)(x)$ _____   _____

3. Multiply $(a \cdot b)(x)$ _____   _____

4. Divide $\left(\dfrac{a}{b}\right)(x)$ _____   _____

5. How do you solve it? $a(2) = 3x + 1$ $c(4) = \dfrac{1}{4}x + 3$

Solve each function.   _____   _____

_____   _____

6. How do you solve it? $b(-2) = 2x - 9$ $b(-2) = 2x - 9$

Solve each function.   _____   _____

_____   _____

Review 1. What is a function? _____   Calculator?
2. What is a vertical line test? _____   yes no
3. How are discrete functions different from continuous ones? _____
4. What is the domain and range of a function? _____
5. How does a function show the 4 operations? _____

Is it a   6.   4, 2   4, 3   2, 7   5, 3   2, 7   3, 4
function?
       Yes, it is a function.   No, it is not a function.   Yes, it is a function.   No, it is not a function.

       7.   9, 5   9, 7   8, 7   6, 2   1, 4   3, 2
       Yes, it is a function.   No, it is not a function.   Yes, it is a function.   No, it is not a function.

Ch 14 Ls 2 Specific Functions 115

_____ #1 #2 ____ / 6   #3 ____ / 7 R ____ / 2 T ____ / 15 _____
        Name   Checker

#1   1. How do you solve a function for a binomial? _____
         2. What if there are 2 seperate functions for the same equation? _____
         _____

   #2   1. How do you solve it? $g(x + 1)$ if $g(x) = 4x + 1$

              What's the answer?   _____

                                        _____

         2. How do you solve it? $g(2) + g(3)$ if $g(x) = 5x - 2$

              What's the answer?   _____

                                        _____

         3. How do you solve it? $-2g(a)$ if $g(x) = 6x + 4$

              What's the answer?   _____

                                        _____

         4. What's the 1st step? $-2g(a) + g(a - 1)$ if $g(x) = x^2 - 3$

              What's the 2nd step?   _____

              What happens next?   _____

              What's the answer?   _____

                                        _____

#3 Solve these problems.   Calculator? yes   no

1. a(x + 4) if a(x) = 5x - 2   b(x - 2) if b(x) = 3x + 2

   _____   _____

   _____   _____

2. c(x + 7) if c(x) = -6x + 1   d(x - 6) if d(x) = -4x + 5

   _____   _____

   _____   _____

3. g(x - 3) if g(x) = 2x + 4   m(x + 8) if m(x) = -5x - 4

   _____   _____

   _____   _____

4.   -2g(a) + g(a - 2) if g(x) = $x^2 - 3x + 8$

   What's the 1st step?   _____

   What's the 2nd step?   _____

   What happens next?   _____

   What's the answer?   _____

Review 1. How do you solve a function for a binomial? _____   Calculator?
       2. What if there are 2 seperate functions for the same equation? _____   yes   no

_____

Ch 14 Ls 3 Piecewise Functions with Quadratics

_____ #1 #2 ____ / 6 #3 ____ / 8 R ____ / 3 T ____ / 17 _____
        Name   Checker

#1 1. What are piecewise functions? _____
       2. How do Piecewise Functions use domains? _____
       3. What is it called when the lines don't meet each other? _____

#2 1. Is there discontinuity? $f(x) = \begin{cases} x + 1 & \text{if } x < 2 \\ 3 & \text{if } x \geq 2 \end{cases}$
        What's the key point?

        Discontinuity   Yes   no

        Key point _____

2. Is there discontinuity? $f(x) = \begin{cases} x^2 + 1 & \text{if } x < 2 \\ 3 & \text{if } x \geq 2 \end{cases}$
   What are the key points?

   Discontinuity   Yes   no

   Key points _____

3. Is there discontinuity? $f(x) = \begin{cases} x^2 - 1 & \text{if } x < 2 \\ 4 & \text{if } x \geq 2 \end{cases}$
   What are the key points?

   Discontinuity   Yes   no

   Key points _____

118

#3 Describe what's happening and a reason it might happen.   Calculator? yes   no

1.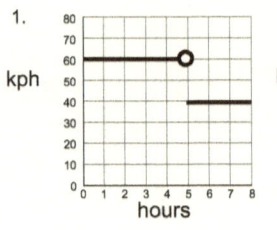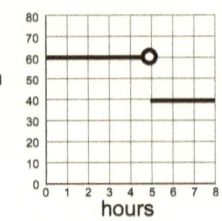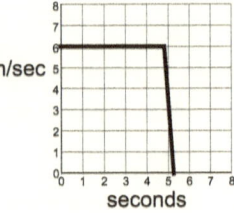

_____   _____   _____

2.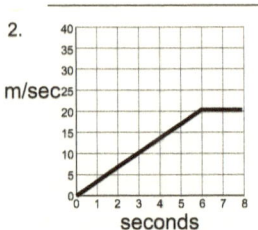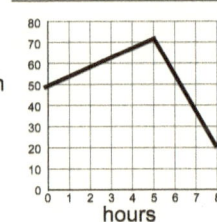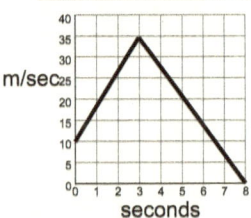

_____   _____   _____

3. Decide how each gas bill got paid.

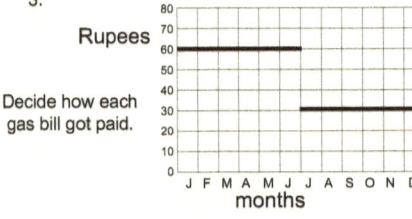   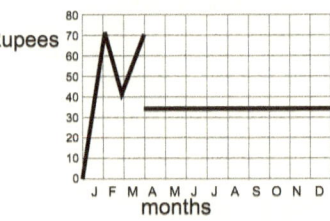

_____   _____
_____   _____

Review 1. How does a graph show you what kind of problem it is? _____   Calculator? yes   no
2. What 2 things shows what the line is doing? _____
3. How does a graph show constant acceleration? _____

Ch 14 Ls 4 Compose Functions 119

_____ #1 #2 ____/ 7 #3 ____/ 12 R ____/ 3 T____/ 22 _____
Name   Checker

#1   1. What does composing functions do? _____

2. Composing functions substitutes... _____

3. What is the order to compose functions? _____

#2 1. Compose these functions. $a(x) = x^2 - 3x$ $b(x) = 2x$
What equation does it make? $(a \circ b)(x)$

Solve for x is 2. What's the answer? _____

2. Compose these functions. $a(x) = x^2 - 3x$ $b(x) = 2x$
What equation does it make? $(b \circ a)(x)$

Solve for x is 2. What's the answer? _____

3. Compose these functions. $a(x) = x^2 + 4x$ $b(x) = 5x$
What equation does it make? $(a \circ b)(x)$

Solve for x is 2. What's the answer? _____

4. Compose these functions. $a(x) = x^2 + 4x$ $b(x) = 5x$
What equation does it make? $(b \circ a)(x)$

Solve for x is 2. What's the answer? _____

$(a \circ b)(x)$

1. $a(x) = x^2 + 4x \quad b(x) = 5x \qquad a(x) = x^2 - 6x \quad b(x) = 3x$ Calculator? yes   no

_____   _____

_____   _____

2. $a(x) = x^2 - 3x \quad b(x) = 7x \qquad a(x) = 2x^2 + 4x \quad b(x) = 9x$

_____   _____

_____   _____

3. $a(x) = 3x^2 + 4x \quad b(x) = 4x \qquad a(x) = 5x^2 + 4x \quad b(x) = -6x$

_____   _____

_____   _____

$(b \circ a)(x)$

4. $a(x) = x^2 + 4x \quad b(x) = 2x \qquad a(x) = x^2 + 8x \quad b(x) = 5x$

_____   _____

_____   _____

5. $a(x) = 2x^2 + 6x \quad b(x) = 9x \qquad a(x) = x^2 + 3x \quad b(x) = 3x$

_____   _____

_____   _____

6. $a(x) = x^2 - 5x \quad b(x) = -7x \qquad a(x) = x^2 - 4x \quad b(x) = -8x$

_____   _____

_____   _____

Review 1. What does composing functions do? _____   Calculator?
2. Composing functions substitutes... _____   yes   no
3. What is the order to compose functions? _____

_____

Review Problems 121

_____ #1 - #3 ____/ 18 Back ____/ 12 Total ____/ 30
Name

1. Piecewise Functions _____

2. Compose Functions _____

#2 Add, Subtract, Multiply, and Divide Functions

$$a(x) = 4x^2 + 6$$
$$b(x) = 2x^2 - 2$$

$$a(x) = -8x^2 + 6$$
$$b(x) = 2x^2 - 1$$

1. Add $(a + b)(x)$ _____   _____

2. Subtract $(a - b)(x)$ _____   _____

3. Multiply $(a \cdot b)(x)$ _____   _____

4. Divide $\left(\dfrac{a}{b}\right)(x)$ _____   _____

5. How do you solve it? $a(7) = x^2 + x - 2$ $c(6) = \dfrac{1}{4}x^2 - 7x$

   Solve each function.   _____   _____

   _____   _____

6. How do you solve it? $b(-2) = 3x^2 - 9x$ $b(-3) = 2x^2 - 8x$

   Solve each function.   _____   _____

   _____   _____

#3 1. $g(x + 7)$ if $g(x) = 6x + 2$ $g(x - 5)$ if $g(x) = 8x + 3$

Solve these   _____   _____
problems.
   _____   _____

   2. $g(x + 4)$ if $g(x) = 7x + 5$ $g(x - 6)$ if $g(x) = 5x - 3$

   _____   _____

   _____   _____

#4 Describe what's happening and a reason it might happen.   Calculator? yes   no

1.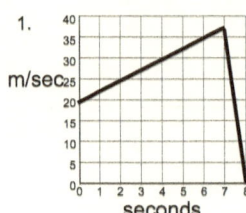
m/sec
seconds

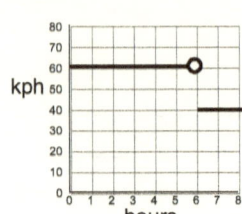
kph
hours

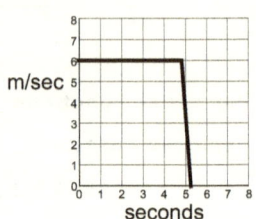
m/sec
seconds

_____   _____   _____
_____   _____   _____

2.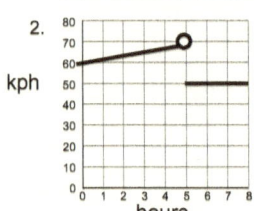
kph
hours

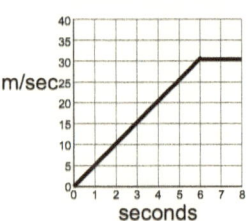
m/sec
seconds

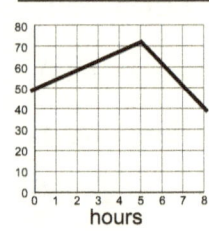
kph
hours

_____   _____   _____
_____   _____   _____

#5 Compose the following functions. $(a \circ b)(x)$ Calculator? yes   no

1. $a(x) = x^2 - 4x$ $b(x) = 5x$ $a(x) = x^2 + 7x$ $b(x) = 9x$

_____   _____
_____   _____

2. $a(x) = x^2 + 9x$ $b(x) = 3x$ $a(x) = 8x^2 - 5x$ $b(x) = 6x$

_____   _____
_____   _____

4. $(b \circ a)(x)$

$a(x) = 3x^2 + 4x$ $b(x) = 2x$ $a(x) = x^2 + 9x$ $b(x) = -5x$

_____   _____
_____   _____

Ch 15 Ls 1 Add and Multiply Inverses. 123

_____ #1 #2 ____ / 9 #3 ____ / 6 R ____ / 5 T ____ / 20 _____
         Name   Checker

#1 1. What does it equal when you add the inverse? _____

2. What does it equal when you multiply the inverse? _____

3. How does - 21 make an additive inverse?   - 21

                                               _____

4. How does 1 fourth make a multiplicative inverse? $\frac{1}{4}$

                                               _____

#2 1. How does a graph find the inverse function of a point? _____

2. Why does the inverse line find the inverse? _____

3. How does X and Y find the inverse function of a point? _____

4. Use the graph to find the inverse.   **(5, 6)**

The inverse is _____.

5. Use the graph to find the inverse.   **(- 3, - 2)**

The inverse is _____.

#3   1. (- 2, - 3) is ___ , ___   (3, - 1) is ___ , ___   Calculator?
                                                                      yes   no

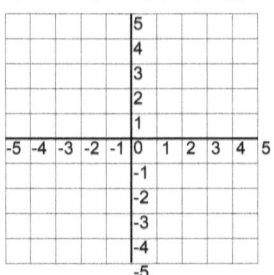   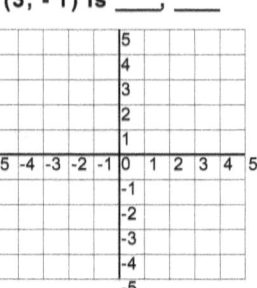

   2. (1, - 2) is ___ , ___   (2, 3) is ___ , ___

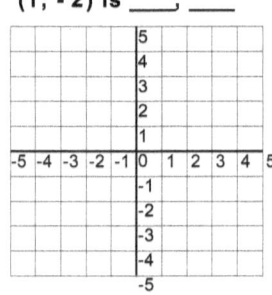   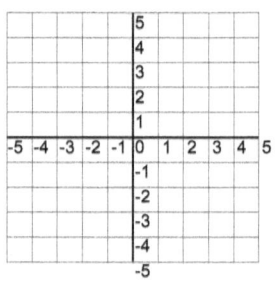

   3. (- 2, 4) is ___ , ___   (- 1, - 3) is ___ , ___

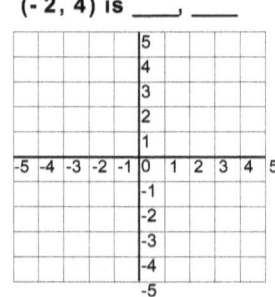   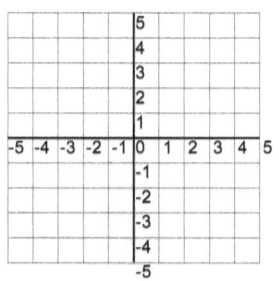

Review 1. What does it equal when you add the inverse? _____   Calculator?
        2. What does it equal when you multiply the inverse? _____   yes   no
        3. How does a graph find the inverse function of a point? _____
        4. Why does the inverse line find the inverse? _____
        5. How does X and Y find the inverse function of a point? _____

Ch 15 Ls 2 Inverse of a Line 125

_____ #1 #2 ____/ 5 #3 ____/ 3 R ____/ 3 T ____/ 11 _____
            Name   Checker

#1 1. How do you find the inverse function of a line? _____

2. What's the 2nd step? _____

3. What 2 steps find standard form? _____

_____

#2 1. First step to find the inverse? $\frac{x}{2} - 2 = y$

Second step?   _____

Last 2 steps?   _____

What's the inverse?   _____

2. First step to find the inverse? $\frac{x}{3} - 5 = y$

Second step?   _____

Last 2 steps?   _____

What's the inverse?   _____

#3 1. y = -0.5x + 2   Calculator? yes no

First step? _____

Second step? _____

Last 2 steps? _____

Inverse is _____

2. y = 2x - 1

First step? _____

Second step? _____

Last 2 steps? _____

Inverse is _____

3. y = 3x + 2

First step? _____

Second step? _____

Last 2 steps? _____

Inverse is _____

Review 1. What's the 1st step to find the inverse of a line? _____   Calculator? yes no

2. What's the 2nd step? _____

3. What 2 steps find standard form? _____

_____

Ch 15 Ls 3 Quadratic Inverses. 127

_____ #1 #2 ____/ 7 #3 ____/ 3 R ____/ 5 T____/ 15 _____
    Name   Checker

#1 1. Why does a parabola not make an inverse function? _____

2. How can you look ahead and tell an inverse is a function? _____

3. How does a quadratic make an inverse function? $y = x^2 + 9$ _____

4. What's the last step to find the inverse function? $\sqrt{x - 9} = y$ _____

5. Why is that step necessary? _____

#2   1. Solve for X is 2 and - 2. $y = -x^2 + 3$

Predict a graph. X intercepts?   When x is 2 it's _____   When x is - 2 it's _____.

Opens Up Down
Y intercept ___

X intercepts
yes   no

2. Solve for X is 2 and - 2. $y = x^2 - 2$

Predict a graph. X intercepts?   When x is 2 it's _____   When x is - 2 it's _____.

Opens Up Down
Y intercept ___

X intercepts
yes   no

#3 1. $y = -x^2 + 3 \quad x \leq 0$

When x is 2 it's _____ When x is -2 it's _____.

Predict a graph. X intercepts?

Find the inverse for the functions.

Opens Up Down

Y intercept ____

X intercepts **yes** **no**

2. $y = x^2 - 2 \quad x \leq 0$

When x is 2 it's _____ When x is -2 it's _____.

Predict a graph. X intercepts?

Opens Up Down

Y intercept ____

X intercepts **yes** **no**

3. $y = x^2 + 4 \quad x \leq 0$

When x is 2 it's _____ When x is -2 it's _____.

Predict a graph. X intercepts?

Opens Up Down

Y intercept ____

X intercepts **yes** **no**

Calculator? yes no

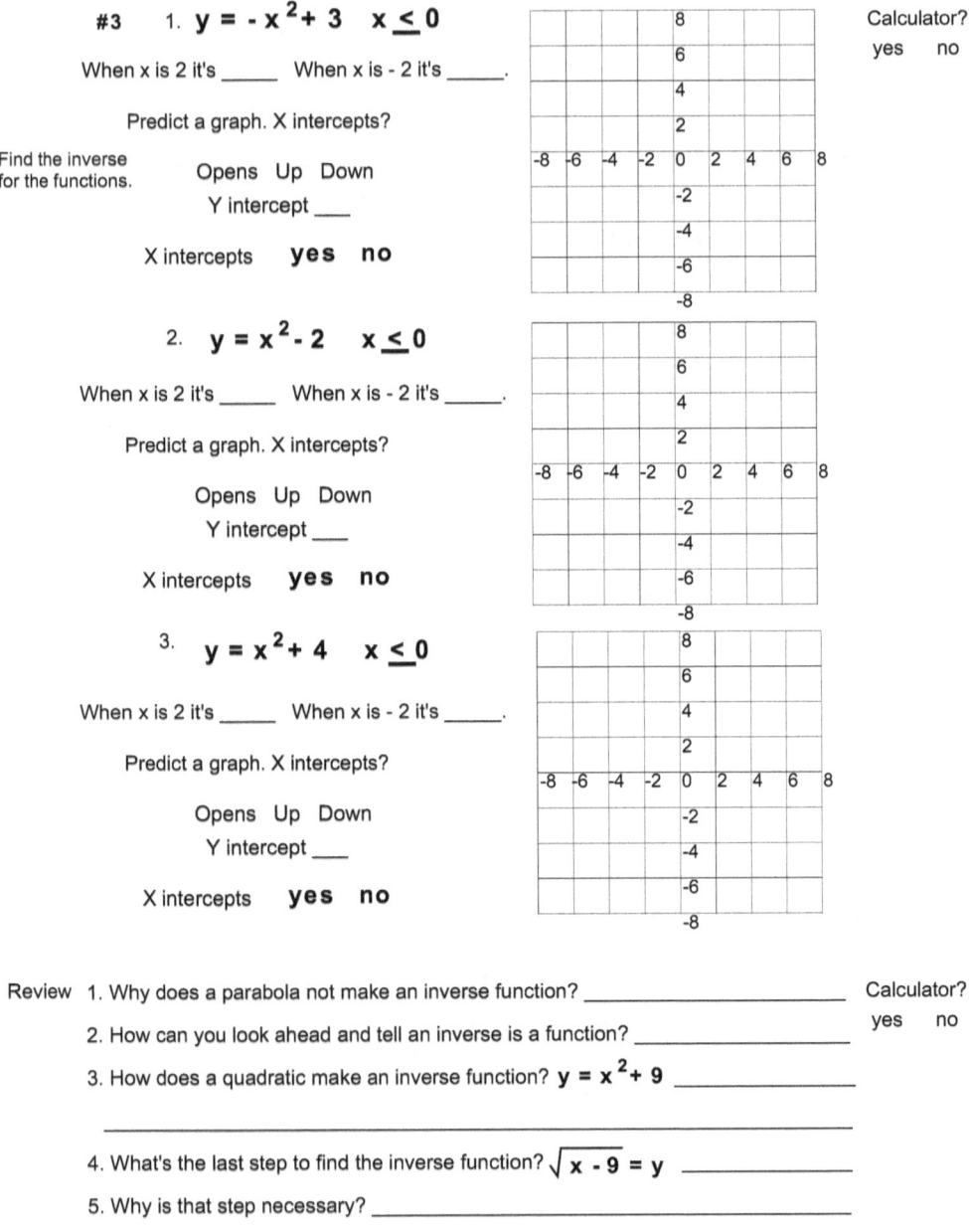

Review 1. Why does a parabola not make an inverse function? _____

2. How can you look ahead and tell an inverse is a function? _____

3. How does a quadratic make an inverse function? $y = x^2 + 9$ _____

_____

4. What's the last step to find the inverse function? $\sqrt{x - 9} = y$ _____

5. Why is that step necessary? _____

Calculator? yes no

Review Problems 129

_____ #1 - #3 ____/ 16 Back ____/ 4 Total ____/ 20
            Name

1. Inverse Functions _____

2. Point Inverse _____

3. Line Inverse _____

4. Quadratic Inverse _____

#2 1.   4 = ____   9 = ____   - 21 = ____   - 25 = ____
Find the   Additive
inverse.
    2. $\frac{1}{4}$ = ____ $\frac{1}{3}$ = ____ $-\frac{1}{2}$ = ____ $-\frac{2}{3}$ = ____
         Multiplicative

3.

(- 1, - 4) is   (2, - 3) is

____ , ____   ____ , ____

Find the
line inverse.   #3   1.   y = 3x - 1

            First step?   _____

            Second step?   _____

            Last 2 steps?   _____

            Inverse is   _____

                    2.   y = x + 2

            First step?   _____

            Second step?   _____

            Last 2 steps?   _____

            Inverse is   _____

Continued   3.. $y = 4x + 2$ Calculator?
                                                              yes   no
First step? _____

Second step? _____

Last 2 steps? _____

Inverse is _____

#4   1. $y = -x^2 + 2$ $x \leq 0$

When x is 2 it's _____   When x is -2 it's _____.

Find the quadratic inverse.   Predict a graph. X intercepts?

Opens Up Down

Y intercept ___

X intercepts   **yes   no**

2. $y = x^2 - 3$ $x \leq 0$

When x is 2 it's _____   When x is -2 it's _____.

Predict a graph. X intercepts?

Opens Up Down

Y intercept ___

X intercepts   **yes   no**

3. $y = x^2 + 4$ $x \leq 0$

When x is 2 it's _____   When x is -2 it's _____.

Predict a graph. X intercepts?

Opens Up Down

Y intercept ___

X intercepts   **yes   no**

Ch 1 Ls 1 What is a matrix?   131

_____ #1 #2 ____/11 #3 ____/ 9 R ___/11 Total ____/ 31 _____
             Name   Checker

#1 1. What is a matrix? _____
   2. What are the numbers going across called? _____
   3. What are the numbers going up or down called? _____
   4. What does a scalar do? _____
   5. What scalar makes a 10% off sale? _____

#2 1. How does a scalar double this matrix?   _____ $\begin{bmatrix} 50 & 70 \\ 40 & 30 \end{bmatrix}$

   Multiply _____. What is the new matrix?   _____ $\begin{bmatrix} 50 & 70 \\ 40 & 30 \end{bmatrix}$

   $\begin{bmatrix} \_\_ & \_\_ \\ \_\_ & \_\_ \end{bmatrix}$

   2. Same matrix equation. How does
      a scalar find 20% off this one?   _____ $\begin{bmatrix} 50 & 70 \\ 40 & 30 \end{bmatrix}$

   Multiply _____. What is the new matrix?   _____ $\begin{bmatrix} 50 & 70 \\ 40 & 30 \end{bmatrix}$

   $\begin{bmatrix} \_\_ & \_\_ \\ \_\_ & \_\_ \end{bmatrix}$

   3. Same matrix. How does a scalar
      find 20% off this matrix?   _____ $\begin{bmatrix} 50 & 70 \\ 40 & 30 \end{bmatrix}$

   Multiply _____. What is the new matrix?   _____ $\begin{bmatrix} 50 & 70 \\ 40 & 30 \end{bmatrix}$

   $\begin{bmatrix} \_\_ & \_\_ \\ \_\_ & \_\_ \end{bmatrix}$

#3 1.

Multiply the scalars.

$1.50 \begin{bmatrix} 2 & 3 \\ 1 & 5 \end{bmatrix}$ $\begin{bmatrix} \_\_ & \_\_ \\ \_\_ & \_\_ \end{bmatrix}$

$2 \begin{bmatrix} 4 & 8 \\ 4 & 9 \end{bmatrix}$ $\begin{bmatrix} \_\_ & \_\_ \\ \_\_ & \_\_ \end{bmatrix}$

$5 \begin{bmatrix} 5 & 3 \\ 7 & 8 \end{bmatrix}$ $\begin{bmatrix} \_\_ & \_\_ \\ \_\_ & \_\_ \end{bmatrix}$

Calculator? yes   no

2. $1.10 \begin{bmatrix} 20 & 30 \\ 10 & 50 \end{bmatrix}$ $\begin{bmatrix} \_\_ & \_\_ \\ \_\_ & \_\_ \end{bmatrix}$

$0.90 \begin{bmatrix} 40 & 30 \\ 60 & 20 \end{bmatrix}$ $\begin{bmatrix} \_\_ & \_\_ \\ \_\_ & \_\_ \end{bmatrix}$

$1.50 \begin{bmatrix} 70 & 20 \\ 90 & 40 \end{bmatrix}$ $\begin{bmatrix} \_\_ & \_\_ \\ \_\_ & \_\_ \end{bmatrix}$

3. $2 \begin{bmatrix} 2.7 & 2.9 \\ 1.5 & 3.2 \end{bmatrix}$ $\begin{bmatrix} \_\_ & \_\_ \\ \_\_ & \_\_ \end{bmatrix}$

$3 \begin{bmatrix} 1.4 & 2.2 \\ 1.7 & 4.5 \end{bmatrix}$ $\begin{bmatrix} \_\_ & \_\_ \\ \_\_ & \_\_ \end{bmatrix}$

$6 \begin{bmatrix} 3.4 & 7.0 \\ 2.3 & 5.6 \end{bmatrix}$ $\begin{bmatrix} \_\_ & \_\_ \\ \_\_ & \_\_ \end{bmatrix}$

Review   1. What is a matrix? _____

2. What are the numbers going across called? _____

3. What are the numbers going up or down called? _____

4. What does a scalar do? _____

5. What scalar makes a 10% off sale? _____

Calculator? yes   no

6. $2.5 \begin{bmatrix} 60 & 40 \\ 20 & 70 \end{bmatrix}$ $\begin{bmatrix} \_\_ & \_\_ \\ \_\_ & \_\_ \end{bmatrix}$

$1.20 \begin{bmatrix} 70 & 50 \\ 60 & 10 \end{bmatrix}$ $\begin{bmatrix} \_\_ & \_\_ \\ \_\_ & \_\_ \end{bmatrix}$

$0.5 \begin{bmatrix} 26 & 38 \\ 42 & 62 \end{bmatrix}$ $\begin{bmatrix} \_\_ & \_\_ \\ \_\_ & \_\_ \end{bmatrix}$

Multiply the scalars.

7. $1.50 \begin{bmatrix} 40 & 60 \\ 70 & 100 \end{bmatrix}$ $\begin{bmatrix} \_\_ & \_\_ \\ \_\_ & \_\_ \end{bmatrix}$

$0.60 \begin{bmatrix} 60 & 40 \\ 80 & 10 \end{bmatrix}$ $\begin{bmatrix} \_\_ & \_\_ \\ \_\_ & \_\_ \end{bmatrix}$

$1.90 \begin{bmatrix} 80 & 90 \\ 50 & 30 \end{bmatrix}$ $\begin{bmatrix} \_\_ & \_\_ \\ \_\_ & \_\_ \end{bmatrix}$

Ch 1 Ls 2 Matrices and Geometry   133

_____ #1 #2 ____ / 9 #3 ____ / 6 R ____ / 9 Total ____ / 24 _____
        Name   Checker

#1 1. How does a matrix show points in geometry? _____
2. What matrix shows the point **3, 2**? _____
3. How is that different from how a slope uses fractions? _____
4. If a matrix has same Y's, what kind of line is it? _____
5. If a matrix has same X's, what kind of line is it? _____
6. How do you solve 2 matrix points for a slope? _____
7. How is a matrix different from slope fractions? _____

#2 1. Name the 2 points. $\begin{bmatrix} 5 & 8 \\ 7 & 2 \end{bmatrix}$

(___, ___) (___, ___)   Describe the graph.
                         What is the slope?

positive   negative
 slope   slope

_____
       Slope

2. Name the 2 points. $\begin{bmatrix} 6 & 5 \\ 3 & 4 \end{bmatrix}$

(___, ___) (___, ___)   Describe the graph.
                         What is the slope?

positive   negative
 slope   slope

_____
       Slope

134.

#3 Name the 2 points.

1. $\begin{bmatrix} 1 & 8 \\ 7 & 2 \end{bmatrix}$ (__, __) (__, __) _____ Calculator? yes   no

Does the 2nd point go up, down, or straight across?

2. $\begin{bmatrix} 9 & 6 \\ 2 & 4 \end{bmatrix}$ (__, __) (__, __) _____

3. $\begin{bmatrix} 1 & 5 \\ 6 & 6 \end{bmatrix}$ (__, __) (__, __) _____

4. $\begin{bmatrix} 2 & 8 \\ 6 & 3 \end{bmatrix}$ (__, __) (__, __) _____

5. $\begin{bmatrix} 4 & 6 \\ 5 & 2 \end{bmatrix}$ (__, __) (__, __) _____

6. $\begin{bmatrix} 2 & 7 \\ 6 & 5 \end{bmatrix}$ (__, __) (__, __) _____

Review 1. How does a matrix show points in geometry? _____   Calculator?
2. What matrix shows the point **3, 2**? _____   yes   no
3. How do you mentally find slope? _____
4. If a matrix has same Y's, what kind of line is it? _____
5. If a matrix has same X's, what kind of line is it? _____
6. How do you solve 2 matrix points for a slope? _____
7. How is a matrix different from slope fractions? _____

8. Make the points and if it positive or negative.

$\begin{bmatrix} 2 & 5 \\ 4 & 4 \end{bmatrix}$

positive
negative
slope

_____ Slope

$\begin{bmatrix} 1 & 4 \\ 1 & 4 \end{bmatrix}$

positive
negative
slope

_____ Slope

Ch 1 Ls 3 How matrices change shapes.   135

_____ #1 #2 ____ /6 #3 ____ / 4 R ___ / 4 Total ____ / 14 _____
           Name   Checker

#1 1. What matrix shows 1 way to make a right triangle? _____
       2. What do you look for to make straight lines? _____
       3. What does it mean to dilate a shape? _____
       4. How does a matrix double the size of a shape? _____

#2 1. Multiply. How does the triangle change? $1.50 \begin{bmatrix} 5 & 8 & 4 \\ 7 & 2 & 4 \end{bmatrix}$

       Describe the graph. What is the shape? $\begin{bmatrix} \_ & \_ & \_ \\ \_ & \_ & \_ \end{bmatrix}$

       _____ Shape

       2. Multiply. How does the triangle change? $0.90 \begin{bmatrix} 1 & 2 & 4 \\ 4 & 2 & 4 \end{bmatrix}$

       Describe the graph. What is the shape? $\begin{bmatrix} \_ & \_ & \_ \\ \_ & \_ & \_ \end{bmatrix}$

       _____ Shape

136.

#3 Multiply. How does the triangle change? Calculator? yes no

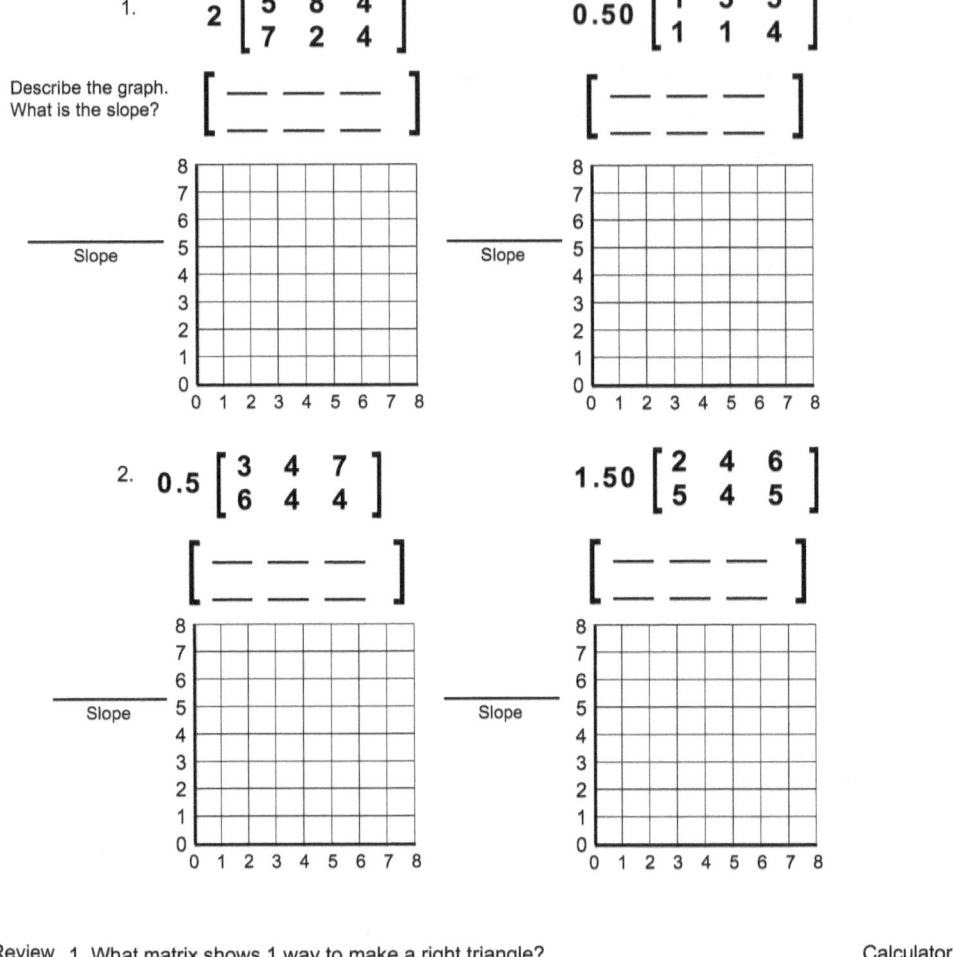

1. $2 \begin{bmatrix} 5 & 8 & 4 \\ 7 & 2 & 4 \end{bmatrix}$

Describe the graph. What is the slope?

$0.50 \begin{bmatrix} 1 & 5 & 5 \\ 1 & 1 & 4 \end{bmatrix}$

2. $0.5 \begin{bmatrix} 3 & 4 & 7 \\ 6 & 4 & 4 \end{bmatrix}$

$1.50 \begin{bmatrix} 2 & 4 & 6 \\ 5 & 4 & 5 \end{bmatrix}$

Review 1. What matrix shows 1 way to make a right triangle? _____   Calculator? yes no
2. What do you look for to make straight lines? _____
3. What does it mean to dilate a shape? _____
4. How does a matrix double the size of a shape? _____

Ch 1 Ls 4 Add matrices. 137

_____ #1 #2 ____/ 8 #3 ____/ 4 R ____/ 4 Total ____/ 16 _____
Name   Checker

#1 1. When can you add matrices? _____

2. How do you add matrices? _____

3. How does a shape move if you add 1 to all the X's in a shape? _____

4. How does a shape move if you subtract 2 from the Y's? _____

#2 1. Add the matrices. $\begin{bmatrix} 1 & 1 \\ 2 & 6 \end{bmatrix} + \begin{bmatrix} 0 & 0 \\ 1 & 1 \end{bmatrix}$

Graph the 1st line.
How does it move? $\begin{bmatrix} \underline{} & \underline{} \\ \underline{} & \underline{} \end{bmatrix}$

The line moves

_____

2. Add the matrices. $\begin{bmatrix} 2 & 5 \\ 2 & 5 \end{bmatrix} + \begin{bmatrix} 2 & 2 \\ 0 & 0 \end{bmatrix}$

Graph the 1st line.
How does it move? $\begin{bmatrix} \underline{} & \underline{} \\ \underline{} & \underline{} \end{bmatrix}$

The line moves

_____

138.

#3 Add the matrices and graph it.   Calculator?
                                                                       yes   no

1. $\begin{bmatrix} 1 & 5 \\ 3 & 3 \end{bmatrix} + \begin{bmatrix} 1 & 1 \\ 0 & 0 \end{bmatrix}$ $\begin{bmatrix} 3 & 5 \\ 3 & 6 \end{bmatrix} + \begin{bmatrix} 0 & 0 \\ 1 & 1 \end{bmatrix}$

Graph the 1st line. $\begin{bmatrix} \_\_ & \_\_ \\ \_\_ & \_\_ \end{bmatrix}$ $\begin{bmatrix} \_\_ & \_\_ \\ \_\_ & \_\_ \end{bmatrix}$
How does it move?

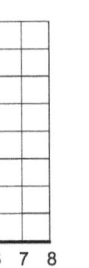   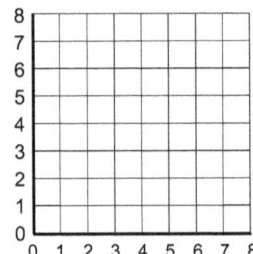

2. $\begin{bmatrix} 2 & 5 \\ 2 & 3 \end{bmatrix} + \begin{bmatrix} -1 & -1 \\ 0 & 0 \end{bmatrix}$ $\begin{bmatrix} 2 & 3 \\ 4 & 6 \end{bmatrix} + \begin{bmatrix} 1 & 1 \\ 1 & 1 \end{bmatrix}$

Graph the 1st line. $\begin{bmatrix} \_\_ & \_\_ \\ \_\_ & \_\_ \end{bmatrix}$ $\begin{bmatrix} \_\_ & \_\_ \\ \_\_ & \_\_ \end{bmatrix}$
How does it move?

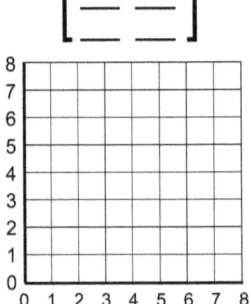

Review 1. When can you add matrices? _____

        2. How do you add matrices? _____

        3. How does a shape move if you add 1 to all the X's in a shape? _____

        4. How does a shape move if you subtract 2 from the Y's? _____

Ch 1 Ls 5 Begin Multiply Matrices. 139

_____ #1 #2 ____/ 10 #3 ____/ 3 R ___/ 13 Total ____/ 26 _____
　　　　Name　　　　　　　　　　　　　　　　　　　　　　　　　　　　　　Checker

#1 1. When can you multiply matrices? _____
   2. Where do you start to multiply matrices? _____
   3. The 1st matrix moves across. How does the 2nd one move? _____
   4. What are the 2 steps to multiply matrices? _____
   _____

#2 1. Can you multiply these matrices? $\begin{bmatrix} 5 & 8 & 9 \\ 7 & 2 & 0 \end{bmatrix} \begin{bmatrix} 5 & 8 \\ 7 & 2 \end{bmatrix}$

　　　　　　　　　　　　　　　　　　　　　　yes　　no

   2. Can you multiply these matrices? $\begin{bmatrix} 5 & 8 \\ 7 & 2 \end{bmatrix} \begin{bmatrix} 5 & 8 & 9 \\ 7 & 2 & 0 \end{bmatrix}$

　　　　　　　　　　　　　　　　　　　　　　yes　　no

   3. Multiply these. What's the answer? $\begin{bmatrix} 5 & 7 \end{bmatrix} \begin{bmatrix} 5 \\ 2 \end{bmatrix}$

　　　　　　　　　　　　　　　　___ + ___ = ___

   4. Multiply these. What's the answer? $\begin{bmatrix} 3 & 3 \end{bmatrix} \begin{bmatrix} 2 \\ 4 \end{bmatrix}$

　　　　　　　　　　　　　　　　___ + ___ = ___

   5. Multiply these. $\begin{bmatrix} 1 & 4 \end{bmatrix} \begin{bmatrix} 2 \\ 2 \end{bmatrix}$

　　　　　　　　　　　　　　　　___ + ___ = ___

   6. Multiply these. $\begin{bmatrix} 6 & 3 \end{bmatrix} \begin{bmatrix} 3 \\ 4 \end{bmatrix}$

　　　　　　　　　　　　　　　　___ + ___ = ___

140.

#3 Can you multiply these matrices? Calculator? yes no

1. $\begin{bmatrix} 5 & 8 & 9 \\ 7 & 2 & 0 \end{bmatrix} \begin{bmatrix} 5 & 8 \\ 7 & 2 \end{bmatrix}$ $\begin{bmatrix} 5 & 8 \\ 7 & 2 \end{bmatrix} \begin{bmatrix} 5 & 8 & 9 \\ 7 & 2 & 0 \end{bmatrix}$

   yes no   yes no

2. $\begin{bmatrix} 5 & 8 & 9 \\ 7 & 2 & 0 \end{bmatrix} \begin{bmatrix} 5 & 8 \\ 7 & 2 \\ 1 & 4 \end{bmatrix}$ $\begin{bmatrix} 5 & 8 \\ 7 & 2 \end{bmatrix} \begin{bmatrix} 5 & 8 & 9 \\ 7 & 2 & 0 \\ 1 & 3 & 1 \end{bmatrix}$

   yes no   yes no

3. $\begin{bmatrix} 4 \\ 7 \end{bmatrix} \begin{bmatrix} 5 & 8 \\ 7 & 2 \end{bmatrix}$ $\begin{bmatrix} 5 & 8 \\ 7 & 2 \end{bmatrix} \begin{bmatrix} 5 \\ 7 \\ 8 \end{bmatrix}$

   yes no   yes no

Review 1. When can you multiply matrices? _____   Calculator? yes no

2. Where do you start to multiply matrices? _____

3. The 1st matrix moves across. How does the 2nd one move? _____

4. What are the 2 steps to multiply matrices? _____
_____

Multiply these matrices.

5. $\begin{bmatrix} 1 & 4 \end{bmatrix} \begin{bmatrix} 2 \\ 7 \end{bmatrix}$ $\begin{bmatrix} 2 & 4 \end{bmatrix} \begin{bmatrix} 2 \\ 2 \end{bmatrix}$ $\begin{bmatrix} 6 & 3 \end{bmatrix} \begin{bmatrix} 1 \\ 2 \end{bmatrix}$

   ___ + ___ = ___   ___ + ___ = ___   ___ + ___ = ___

6. $\begin{bmatrix} 6 & 1 \end{bmatrix} \begin{bmatrix} 3 \\ 2 \end{bmatrix}$ $\begin{bmatrix} 2 & 1 \end{bmatrix} \begin{bmatrix} 3 \\ 3 \end{bmatrix}$ $\begin{bmatrix} 4 & 2 \end{bmatrix} \begin{bmatrix} 4 \\ 6 \end{bmatrix}$

   ___ + ___ = ___   ___ + ___ = ___   ___ + ___ = ___

7. $\begin{bmatrix} 2 & 8 \end{bmatrix} \begin{bmatrix} 2 \\ 2 \end{bmatrix}$ $\begin{bmatrix} 4 & 4 \end{bmatrix} \begin{bmatrix} 2 \\ 5 \end{bmatrix}$ $\begin{bmatrix} 5 & 7 \end{bmatrix} \begin{bmatrix} 5 \\ 5 \end{bmatrix}$

   ___ + ___ = ___   ___ + ___ = ___   ___ + ___ = ___

Ch 1 Ls 6 Multiply Larger Matrices   141

_____ #1 #2 ____ / 9 #3 ____ / 8 R ____ / 5 Total ____ / 22 _____
Name   Checker

#1 1. Where do larger matrices start to multiply? _____

2. What are the 2 steps to multiply matrices? _____

3. What multiplies next if the 1st matrix has more rows? _____

4. What does the left matrix control? _____

5. What does the right matrix control? _____

#2 1. Start at the 1st element and multiply the 1st row. $\begin{bmatrix} 5 & 8 \\ 7 & 2 \end{bmatrix} \begin{bmatrix} 5 \\ 7 \end{bmatrix}$

Multiply the 2nd row. ___ + ___ = $\begin{bmatrix} \underline{} \\ \end{bmatrix}$

___ + ___ = $\begin{bmatrix} \\ \underline{} \end{bmatrix}$

2. Start at the 1st element and multiply the 1st row. $\begin{bmatrix} 1 & 4 \\ 3 & 2 \end{bmatrix} \begin{bmatrix} 1 \\ 2 \end{bmatrix}$

Multiply the 2nd row. ___ + ___ = $\begin{bmatrix} \underline{} \\ \end{bmatrix}$

___ + ___ = $\begin{bmatrix} \\ \underline{} \end{bmatrix}$

3. Start at the 1st element and multiply the 1st row. $\begin{bmatrix} 3 & 6 \\ 3 & 2 \end{bmatrix} \begin{bmatrix} 4 \\ 2 \end{bmatrix}$

Multiply the 2nd row. ___ + ___ = $\begin{bmatrix} \underline{} \\ \end{bmatrix}$

___ + ___ = $\begin{bmatrix} \\ \underline{} \end{bmatrix}$

4. Multiply all 4 steps. What's the answer? $\begin{bmatrix} 5 & 8 \\ 5 & 4 \end{bmatrix} \begin{bmatrix} 1 & 1 \\ 4 & 2 \end{bmatrix}$ Add these mentally.

$\begin{bmatrix} \underline{} & \underline{} \\ \underline{} & \underline{} \end{bmatrix}$

#3 1. $\begin{bmatrix} 1 & 4 \\ 1 & 2 \end{bmatrix} \begin{bmatrix} 2 \\ 6 \end{bmatrix}$ $\begin{bmatrix} 2 & 8 \\ 3 & 2 \end{bmatrix} \begin{bmatrix} 1 \\ 4 \end{bmatrix}$ Calculator? yes   no

Multiply these matrices.

___ + ___ = [___]
___ + ___ = [___]

___ + ___ = [___]
___ + ___ = [___]

2. $\begin{bmatrix} 4 & 6 \\ 5 & 3 \end{bmatrix} \begin{bmatrix} 2 \\ 8 \end{bmatrix}$ $\begin{bmatrix} 3 & 5 \\ 1 & 1 \end{bmatrix} \begin{bmatrix} 3 \\ 3 \end{bmatrix}$

___ + ___ = [___]
___ + ___ = [___]

___ + ___ = [___]
___ + ___ = [___]

3. $\begin{bmatrix} 2 & 1 \\ 7 & 2 \end{bmatrix} \begin{bmatrix} 4 & 6 \\ 4 & 2 \end{bmatrix}$ $\begin{bmatrix} 4 & 6 \\ 4 & 2 \end{bmatrix} \begin{bmatrix} 2 & 1 \\ 7 & 2 \end{bmatrix}$

Add these mentally.

$\begin{bmatrix} \underline{} & \underline{} \\ \underline{} & \underline{} \end{bmatrix}$ $\begin{bmatrix} \underline{} & \underline{} \\ \underline{} & \underline{} \end{bmatrix}$

4. $\begin{bmatrix} 4 & 5 \\ 3 & 1 \end{bmatrix} \begin{bmatrix} 2 & 3 \\ 1 & 1 \end{bmatrix}$ $\begin{bmatrix} 3 & 3 \\ 1 & 2 \end{bmatrix} \begin{bmatrix} 2 & 5 \\ 2 & 1 \end{bmatrix}$

$\begin{bmatrix} \underline{} & \underline{} \\ \underline{} & \underline{} \end{bmatrix}$ $\begin{bmatrix} \underline{} & \underline{} \\ \underline{} & \underline{} \end{bmatrix}$

Review   1. Where do larger matrices start to multiply? _____

2. What are the 2 steps to multiply matrices? _____

3. What multiplies next if the 1st matrix has more rows? _____

4. What does the left matrix control? _____

5. What does the right matrix control? _____

Review Problems 143

_____ #1 - #4 ____/ 17   Back ____/ 16   Total ____/ 33
              Name

1. Matrix _____
2. Scalar _____
3. When to Add Matrices _____
4. When to Multiply Matrices _____
5. Larger Matrices _____

#2 Multiply the scalars.

1.

$1.50 \begin{bmatrix} 2 & 3 \\ 1 & 5 \end{bmatrix}$
$\begin{bmatrix} \_ & \_ \\ \_ & \_ \end{bmatrix}$

$2 \begin{bmatrix} 4 & 8 \\ 4 & 9 \end{bmatrix}$
$\begin{bmatrix} \_ & \_ \\ \_ & \_ \end{bmatrix}$

$5 \begin{bmatrix} 5 & 3 \\ 7 & 8 \end{bmatrix}$
$\begin{bmatrix} \_ & \_ \\ \_ & \_ \end{bmatrix}$

2.

$1.10 \begin{bmatrix} 20 & 30 \\ 10 & 50 \end{bmatrix}$
$\begin{bmatrix} \_ & \_ \\ \_ & \_ \end{bmatrix}$

$0.90 \begin{bmatrix} 40 & 30 \\ 60 & 20 \end{bmatrix}$
$\begin{bmatrix} \_ & \_ \\ \_ & \_ \end{bmatrix}$

$1.50 \begin{bmatrix} 70 & 20 \\ 90 & 40 \end{bmatrix}$
$\begin{bmatrix} \_ & \_ \\ \_ & \_ \end{bmatrix}$

#3 Does the 2nd point go up, down, or straight across?

1. $\begin{bmatrix} 1 & 8 \\ 7 & 2 \end{bmatrix}$ (__, __) (__, __)
   Up Down Straight

   $\begin{bmatrix} 1 & 5 \\ 6 & 6 \end{bmatrix}$ (__, __) (__, __)
   Up Down Straight

2. $\begin{bmatrix} 9 & 6 \\ 2 & 4 \end{bmatrix}$ (__, __) (__, __)
   Up Down Straight

   $\begin{bmatrix} 1 & 4 \\ 6 & 5 \end{bmatrix}$ (__, __) (__, __)
   Up Down Straight

#4 Multiply. How does the triangle change?

1. $2 \begin{bmatrix} 5 & 8 & 4 \\ 7 & 2 & 4 \end{bmatrix}$
$\begin{bmatrix} \_ & \_ & \_ \\ \_ & \_ & \_ \end{bmatrix}$

$0.50 \begin{bmatrix} 1 & 5 & 5 \\ 1 & 1 & 4 \end{bmatrix}$
$\begin{bmatrix} \_ & \_ & \_ \\ \_ & \_ & \_ \end{bmatrix}$

_____   _____
_____   _____

#5 Add the matrices.   Calculator? yes no

1. $\begin{bmatrix} 1 & 5 \\ 3 & 3 \end{bmatrix} + \begin{bmatrix} 1 & 1 \\ 0 & 0 \end{bmatrix} = \begin{bmatrix} \underline{} & \underline{} \\ \underline{} & \underline{} \end{bmatrix}$ $\begin{bmatrix} 3 & 5 \\ 3 & 6 \end{bmatrix} + \begin{bmatrix} 0 & 0 \\ 1 & 1 \end{bmatrix} = \begin{bmatrix} \underline{} & \underline{} \\ \underline{} & \underline{} \end{bmatrix}$

2. $\begin{bmatrix} 0 & 5 \\ 4 & -2 \end{bmatrix} + \begin{bmatrix} 3 & 3 \\ -1 & -1 \end{bmatrix} = \begin{bmatrix} \underline{} & \underline{} \\ \underline{} & \underline{} \end{bmatrix}$ $\begin{bmatrix} -2 & 2 \\ 4 & -3 \end{bmatrix} + \begin{bmatrix} -2 & -2 \\ 4 & 4 \end{bmatrix} = \begin{bmatrix} \underline{} & \underline{} \\ \underline{} & \underline{} \end{bmatrix}$

3. $\begin{bmatrix} -7 & 2 \\ -1 & 4 \end{bmatrix} + \begin{bmatrix} 6 & 6 \\ -2 & -2 \end{bmatrix} = \begin{bmatrix} \underline{} & \underline{} \\ \underline{} & \underline{} \end{bmatrix}$ $\begin{bmatrix} 4 & 9 \\ 2 & 7 \end{bmatrix} + \begin{bmatrix} -4 & -4 \\ 5 & 5 \end{bmatrix} = \begin{bmatrix} \underline{} & \underline{} \\ \underline{} & \underline{} \end{bmatrix}$

#6 Multiply these matrices.   Calculator? yes no

1. $\begin{bmatrix} 1 & 4 \\ 1 & 2 \end{bmatrix} \begin{bmatrix} 2 \\ 6 \end{bmatrix}$ $\underline{} * \underline{} + \underline{} * \underline{} = \begin{bmatrix} \underline{} \\ \underline{} \end{bmatrix}$

2. $\begin{bmatrix} 4 & 6 \\ 5 & 3 \end{bmatrix} \begin{bmatrix} 2 \\ 8 \end{bmatrix}$ $\underline{} * \underline{} + \underline{} * \underline{} = \begin{bmatrix} \underline{} \\ \underline{} \end{bmatrix}$

3. $\begin{bmatrix} 2 & 8 \\ 3 & 2 \end{bmatrix} \begin{bmatrix} 1 \\ 4 \end{bmatrix}$ $\underline{} * \underline{} + \underline{} * \underline{} = \begin{bmatrix} \underline{} \\ \underline{} \end{bmatrix}$

4. $\begin{bmatrix} 3 & 5 \\ 1 & 1 \end{bmatrix} \begin{bmatrix} 3 \\ 3 \end{bmatrix}$ $\underline{} * \underline{} + \underline{} * \underline{} = \begin{bmatrix} \underline{} \\ \underline{} \end{bmatrix}$

5. $\begin{bmatrix} 2 & 1 \\ 7 & 2 \end{bmatrix} \begin{bmatrix} 4 & 6 \\ 4 & 2 \end{bmatrix} \begin{bmatrix} \underline{} & \underline{} \\ \underline{} & \underline{} \end{bmatrix}$ $\begin{bmatrix} 4 & 6 \\ 4 & 2 \end{bmatrix} \begin{bmatrix} 2 & 1 \\ 7 & 2 \end{bmatrix} \begin{bmatrix} \underline{} & \underline{} \\ \underline{} & \underline{} \end{bmatrix}$

6. $\begin{bmatrix} 4 & 5 \\ 3 & 1 \end{bmatrix} \begin{bmatrix} 2 & 3 \\ 1 & 1 \end{bmatrix} \begin{bmatrix} \underline{} & \underline{} \\ \underline{} & \underline{} \end{bmatrix}$ $\begin{bmatrix} 3 & 3 \\ 1 & 2 \end{bmatrix} \begin{bmatrix} 2 & 5 \\ 2 & 1 \end{bmatrix} \begin{bmatrix} \underline{} & \underline{} \\ \underline{} & \underline{} \end{bmatrix}$

7. $\begin{bmatrix} 8 & 6 \\ -3 & -4 \end{bmatrix} \begin{bmatrix} 1 & 2 \\ 2 & 3 \end{bmatrix} \begin{bmatrix} \underline{} & \underline{} \\ \underline{} & \underline{} \end{bmatrix}$ $\begin{bmatrix} 5 & 7 \\ 9 & 3 \end{bmatrix} \begin{bmatrix} -2 & 4 \\ -3 & 2 \end{bmatrix} \begin{bmatrix} \underline{} & \underline{} \\ \underline{} & \underline{} \end{bmatrix}$

8. $\begin{bmatrix} -7 & -3 \\ -2 & -1 \end{bmatrix} \begin{bmatrix} -4 & -5 \\ -6 & -2 \end{bmatrix} \begin{bmatrix} \underline{} & \underline{} \\ \underline{} & \underline{} \end{bmatrix}$ $\begin{bmatrix} 2 & 4 \\ 3 & 2 \end{bmatrix} \begin{bmatrix} 5 & 7 \\ 9 & 3 \end{bmatrix} \begin{bmatrix} \underline{} & \underline{} \\ \underline{} & \underline{} \end{bmatrix}$

Ch 2 Ls 1 Matrices as Equations

_____ #1 #2 ____/13 #3 ____/ 6 R ____/13 Total ____/ 32 _____
Name   Checker

#1 1. How do Variables Inside write the equation x + y is 6? _____

2. What kind of matrix does Variables Outside make? _____

3. How does dotted line write the equation 2x + 4y = 5 _____

4. Which way does this matrix use to write equations? $\begin{bmatrix} 4x \\ 4x + 3y \end{bmatrix} = \begin{bmatrix} y \\ 16 \end{bmatrix}$

_____

5. How does this matrix make 2 equations? $\begin{bmatrix} 3 & 2 \\ 4 & -2 \end{bmatrix} \begin{bmatrix} x \\ y \end{bmatrix} = \begin{bmatrix} 1 \\ 6 \end{bmatrix}$

_____

6. What are these equations? $\begin{bmatrix} 1 & 3 & \vdots & -5 \\ 2 & 5 & \vdots & 9 \\ 4 & -2 & \vdots & 7 \end{bmatrix}$

_____   _____   _____

#2 1. Name the 1st 3 ways to change matrices. _____

2. Name the last 2 ways to change matrices. _____

Decide how it changed.   3. $\begin{bmatrix} 4 & 7 \\ 6 & -8 \end{bmatrix} \begin{bmatrix} 2 \\ 2 \end{bmatrix}$ $\begin{bmatrix} 6 & -8 \\ 4 & 7 \end{bmatrix} \begin{bmatrix} 2 \\ 2 \end{bmatrix}$ $\begin{bmatrix} 2 & 3 \\ 1 & -2 \end{bmatrix} \begin{bmatrix} 1 \\ 7 \end{bmatrix}$ $\begin{bmatrix} 2 & 3 \\ -1 & 2 \end{bmatrix} \begin{bmatrix} 1 \\ -7 \end{bmatrix}$

_____   _____

4. $\begin{bmatrix} 4 & 7 \\ 6 & -8 \end{bmatrix} \begin{bmatrix} 2 \\ 2 \end{bmatrix}$ $\begin{bmatrix} 8 & 14 \\ 6 & -8 \end{bmatrix} \begin{bmatrix} 4 \\ 2 \end{bmatrix}$ $\begin{bmatrix} 2 & 3 \\ 1 & -2 \end{bmatrix} \begin{bmatrix} 1 \\ 7 \end{bmatrix}$ $\begin{bmatrix} 2 & 3 \\ 3 & 1 \end{bmatrix} \begin{bmatrix} 1 \\ 8 \end{bmatrix}$

_____   _____

5. $\begin{bmatrix} 4 & 7 \\ 6 & -8 \end{bmatrix} \begin{bmatrix} 2 \\ 2 \end{bmatrix}$ $\begin{bmatrix} 5 & 4 \\ 1 & -2 \end{bmatrix} \begin{bmatrix} 9 \\ 7 \end{bmatrix}$

_____

146.

#3 Write equations for these matrices.   Calculator?

1. $\begin{bmatrix} 3 & 2 \\ 4 & -2 \end{bmatrix}\begin{bmatrix} x \\ y \end{bmatrix}=\begin{bmatrix} 1 \\ 6 \end{bmatrix}$ $\begin{bmatrix} 1 & 5 \\ 2 & 3 \end{bmatrix}\begin{bmatrix} x \\ y \end{bmatrix}=\begin{bmatrix} 1 \\ 4 \end{bmatrix}$ $\begin{bmatrix} 5x \\ 2x+2y \end{bmatrix}=\begin{bmatrix} y \\ 12 \end{bmatrix}$ yes   no

_____   _____   _____

2. $\begin{bmatrix} 4x \\ 4x+3y \end{bmatrix}=\begin{bmatrix} y \\ 16 \end{bmatrix}$ $\begin{bmatrix} 5 & 4 & | & 5 \\ 3 & 7 & | & 10 \end{bmatrix}$ $\begin{bmatrix} 1 & 3 & | & -5 \\ 2 & 5 & | & 9 \end{bmatrix}$

_____   _____   _____

Review 1. How does Variables Inside write the equation x + y is 6? _____   Calculator?
2. What kind of matrix does Variables Outside make? _____   yes   no
3. How does dotted line write the equation 2x + 4y = 5 _____
4. What are the 1st 3 ways to augment matrices? _____
_____
5. What are the last 2 ways to change them? _____

Decide how it changed.

6. $\begin{bmatrix} 2 & 3 \\ 1 & -2 \end{bmatrix}\begin{bmatrix} 1 \\ 7 \end{bmatrix}$ $\begin{bmatrix} 1 & -2 \\ 2 & 3 \end{bmatrix}\begin{bmatrix} 7 \\ 1 \end{bmatrix}$ $\begin{bmatrix} 2 & 3 \\ 1 & -2 \end{bmatrix}\begin{bmatrix} 1 \\ 7 \end{bmatrix}$ $\begin{bmatrix} 2 & 3 \\ -1 & 2 \end{bmatrix}\begin{bmatrix} 1 \\ -7 \end{bmatrix}$

Multiply Add Multiply/Add Flip Switch   Multiply Add Multiply/Add Flip Switch

7. $\begin{bmatrix} 2 & 3 \\ 1 & -2 \end{bmatrix}\begin{bmatrix} 1 \\ 7 \end{bmatrix}$ $\begin{bmatrix} 4 & 6 \\ 1 & -2 \end{bmatrix}\begin{bmatrix} 2 \\ 7 \end{bmatrix}$ $\begin{bmatrix} 2 & 3 \\ 1 & -2 \end{bmatrix}\begin{bmatrix} 1 \\ 7 \end{bmatrix}$ $\begin{bmatrix} 2 & 3 \\ 3 & 1 \end{bmatrix}\begin{bmatrix} 1 \\ 8 \end{bmatrix}$

Multiply Add Multiply/Add Flip Switch   Multiply Add Multiply/Add Flip Switch

8. $\begin{bmatrix} 2 & 3 \\ 1 & -2 \end{bmatrix}\begin{bmatrix} 1 \\ 7 \end{bmatrix}$ $\begin{bmatrix} 5 & 4 \\ 1 & -2 \end{bmatrix}\begin{bmatrix} 9 \\ 7 \end{bmatrix}$ $\begin{bmatrix} 1 & 4 \\ 5 & 2 \end{bmatrix}\begin{bmatrix} 2 \\ 8 \end{bmatrix}$ $\begin{bmatrix} 5 & 2 \\ 1 & 4 \end{bmatrix}\begin{bmatrix} 8 \\ 2 \end{bmatrix}$

Multiply Add Multiply/Add Flip Switch   Multiply Add Multiply/Add Flip Switch

9. $\begin{bmatrix} 1 & 4 \\ 5 & 2 \end{bmatrix}\begin{bmatrix} 2 \\ 8 \end{bmatrix}$ $\begin{bmatrix} 1 & 4 \\ -1 & 2 \end{bmatrix}\begin{bmatrix} 2 \\ -7 \end{bmatrix}$ $\begin{bmatrix} 1 & 4 \\ 5 & 2 \end{bmatrix}\begin{bmatrix} 2 \\ 8 \end{bmatrix}$ $\begin{bmatrix} 1 & 4 \\ 6 & 6 \end{bmatrix}\begin{bmatrix} 2 \\ 10 \end{bmatrix}$

Multiply Add Multiply/Add Flip Switch   Multiply Add Multiply/Add Flip Switch

Ch 2 Ls 2 Cramer's Law   147

_____ #1 #2 ____/10 #3 ____/ 4   R ____/ 6 Total ____/ 20 _____
          Name   Checker

#1 1. How do you cross multiply Cramer's? _____
   2. What do you do with the determinant to cross multiply? _____
   3. What's the 2nd step to multiply it? _____
   4. What's the 3rd step to multiply it? _____

#2   1. What is the 1st step to cross multiply? $\begin{bmatrix} 4 & 1 \\ 3 & 5 \end{bmatrix}$

Cross multiply. ____ - ____ = ____ $\begin{bmatrix} 4 & 1 \\ 3 & 5 \end{bmatrix}$ What is the next step?

Cross multiply. ____ - ____ = ____ $\begin{bmatrix} 4 & 1 \\ 3 & 5 \end{bmatrix}$

2. What is the 1st step to multiply Cramer's? $\begin{bmatrix} 2 & 3 \\ 4 & 7 \end{bmatrix} = \begin{bmatrix} 2 \\ 8 \end{bmatrix}$

Cross multiply. ____ - ____ = ____ $\begin{bmatrix} 2 & 3 \\ 4 & 7 \end{bmatrix}$ Next.

What is the 2nd step to multiply Cramer's? $\begin{bmatrix} 2 & 3 \\ 4 & 7 \end{bmatrix} = \begin{bmatrix} 2 \\ 8 \end{bmatrix}$

Put C in for X. $\begin{bmatrix} \underline{} & 3 \\ \underline{} & 7 \end{bmatrix}$ Find the answer.

Cross multiply. ____ - ____ = ____   Next.

What's the 3rd step? $\begin{bmatrix} 2 & 3 \\ 4 & 7 \end{bmatrix} = \begin{bmatrix} 2 \\ 8 \end{bmatrix}$

Put C in for Y. $\begin{bmatrix} 2 & \underline{} \\ 4 & \underline{} \end{bmatrix}$ Find the answer.

Cross multiply. ____ - ____ = ____   What are both points?

____   ____

148.

#3 1. What is the 1st step to multiply Cramer's? $\begin{bmatrix} 5 & 6 \\ 4 & 2 \end{bmatrix} = \begin{bmatrix} 3 \\ 9 \end{bmatrix}$ Calculator? yes no

Cross multiply. ____ - ____ = ____ $\begin{bmatrix} 5 & 6 \\ 4 & 2 \end{bmatrix}$ What is the next step?

Use Cramer's to solve these.

Put C in for X. $\begin{bmatrix} \underline{} & 6 \\ & 2 \end{bmatrix}$

What's the 3rd step to change the matrix? $\begin{bmatrix} 5 & 6 \\ 4 & 2 \end{bmatrix} = \begin{bmatrix} 3 \\ 9 \end{bmatrix}$

Put C in for Y. $\begin{bmatrix} 5 & \underline{} \\ 4 & \underline{} \end{bmatrix}$ What are both points?

____ ____

1. What is the 1st step to multiply Cramer's? $\begin{bmatrix} 8 & 2 \\ 5 & 3 \end{bmatrix} = \begin{bmatrix} 5 \\ 6 \end{bmatrix}$

Cross multiply. ____ - ____ = ____ $\begin{bmatrix} 8 & 2 \\ 5 & 3 \end{bmatrix}$ What is the next step?

Put C in for X. $\begin{bmatrix} \underline{} & 2 \\ & 3 \end{bmatrix}$

What's the 3rd step to change the matrix? $\begin{bmatrix} 8 & 2 \\ 5 & 3 \end{bmatrix} = \begin{bmatrix} 5 \\ 6 \end{bmatrix}$

Put C in for Y. $\begin{bmatrix} 8 & \underline{} \\ 5 & \underline{} \end{bmatrix}$ What are both points?

____ ____

Review 1. How do you cross multiply Cramer's? _____   Calculator? yes no

2. What do you do with the determinant to cross multiply? _____

3. What's the 3rd step to multiply it? _____

4. What's the 4th step to multiply it? _____

Ch 2 Ls 3 Cramer's Rule   149

_____ #1 #2 ____/ 8 #3 ____/ 6 R ___/ 4 Total ____/ 18 _____
Name   Checker

#1 1. What does the 1st step of Cramer's rule solve to get? _____
2. What did the 1st step find? _____
3. How do you find the X numerator? _____
4. How do you find the Y numerator? _____

#2 1. What solves the 1st step? $2x + 2y = 3$
                                            $x + 3y = 4$

What's the 2nd step?   ___ ___   ___ · ___ = ___   ⬚ , ⬚
                        ___ ___

What's the last step?   ___ ___   ___ · ___ = ___   ⬚ , ⬚
                        ___ ___

                        ___ ___   ___ · ___ = ___   ⬚ , ⬚
                        ___ ___

2. Solve the 3 steps. What's the answer? $5x + 4y = 5$
                                             $6x + y = 8$

___ · ___ = ___   ___ · ___ = ___   ___ · ___ = ___   ⬚ , ⬚
  1st step   2nd step   last step

3. Solve the 3 steps. What's the answer? $-9x - 3y = 4$
                                             $7x + 2y = 6$

___ · ___ = ___   ___ · ___ = ___   ___ · ___ = ___   ⬚ , ⬚
  1st step   2nd step   last step

4. Solve the 3 steps. What's the answer? $2x - 2y = 6$
                                             $3x - 7y = 8$

___ · ___ = ___   ___ · ___ = ___   ___ · ___ = ___   ⬚ , ⬚
  1st step   2nd step   last step

150.

#3 Solve the 3 steps for Cramer's Law. What's the answer?   Calculator? yes   no

1. $7x + y = 5$
    $9x + 4y = 6$

    1st step ___ · ___ = ___
    2nd step ___ · ___ = ___
    last step ___ · ___ = ___

    ☐ , ☐

2. $2x + 2y = 3$
    $x + 3y = 4$

    1st step ___ · ___ = ___
    2nd step ___ · ___ = ___
    last step ___ · ___ = ___

    ☐ , ☐

3. $8x - 2y = 4$
    $-x + 5y = 6$

    1st step ___ · ___ = ___
    2nd step ___ · ___ = ___
    last step ___ · ___ = ___

    ☐ , ☐

4. $-2x + 5y = 8$
    $6x - 3y = 5$

    1st step ___ · ___ = ___
    2nd step ___ · ___ = ___
    last step ___ · ___ = ___

    ☐ , ☐

5. $-9x - 2y = 3$
    $7x + 4y = 9$

    1st step ___ · ___ = ___
    2nd step ___ · ___ = ___
    last step ___ · ___ = ___

    ☐ , ☐

6. $6x - 3y = 6$
    $x - 4y = 5$

    1st step ___ · ___ = ___
    2nd step ___ · ___ = ___
    last step ___ · ___ = ___

    ☐ , ☐

Review 1. What does the 1st step of Cramer's rule solve to get? _____   Calculator? yes   no

2. How does it solve the 4 numbers? _____

3. After the left side, what's next? _____

4. What finishes the answer? _____

Ch 2 Ls 4 Inverse of Matrices   151

_____ #1 #2 ____/ 6 #3 ____/ 4 R ____/ 3 Total ____/ 13 _____
        Name   Checker

#1 1. How do you eliminate the 3rd variable? _____
    2. How do you finish eliminating the 1st variable? _____
    3. What happens next? _____
    4. How does a 3rd variable change a graph? _____

#2 1. What solves the 1st step with 3 variables? $2x + 2y + z = 3$
                                                          $x + 3y + z = 4$
                                                          $x + 3y + z = 4$

Eliminate _____
            What happens next?   _____

Eliminate _____
            What happens next?   _____

            What's the answer?
                                  _____

                                  _____

    2. What solves the 1st step with 3 variables? $3x + 4y + 2z = 10$
                                                      $x + 8y + z = 8$
                                                      $3x + 3y + 2z = 6$

Eliminate _____
            What happens next?   _____

Eliminate _____
            What happens next?   _____

            What's the answer?
                                  _____

                                  _____

#3 Solve these with Cramer's Law.   Calculator?
                                                                   yes   no

$$4x + 2y + 5z = 6$$
1. What solves the 1st step with 3 variables? $$2x + 3y + 5z = 2$$
$$x + 6y + z = 8$$

Eliminate _____
    What happens next?   _____

Eliminate _____
    What happens next?   _____

    What's the answer?   _____

                                                      _____

$$6x + 4y + 3z = 7$$
2. What solves the 1st step with 3 variables? $$x + 5y + 3z = 2$$
$$x + 10y + z = 4$$

Eliminate _____
    What happens next?   _____

Eliminate _____
    What happens next?   _____

    What's the answer?   _____

                                                      _____

Review 1. How do you eliminate the 3rd variable? _____   Calculator?
      2. How do you finish eliminating the 1st variable? _____   yes   no
      3. How does a 3rd variable change a graph? _____

Review Problems 153

_____ #1 - #3 ____ / 18   Back ____ / 6   Total ____ / 24
                Name

1. 3 Matrix as Equations _____
2. Augmented Matrices _____
3. Cramer's Rule _____
4. 3 Variable Equations _____

#2 1. $\begin{bmatrix} 1 & 2 \\ 6 & -7 \end{bmatrix} \begin{bmatrix} x \\ y \end{bmatrix} = \begin{bmatrix} 5 \\ 8 \end{bmatrix}$ $\begin{bmatrix} 4 & 1 & \vdots & 12 \\ 5 & 8 & \vdots & 20 \end{bmatrix}$ $\begin{bmatrix} 8x - 2y \\ 9x + 3y \end{bmatrix} = \begin{bmatrix} 6 \\ 2 \end{bmatrix}$
Write the
equations.

_____   _____   _____

2. $\begin{bmatrix} 2x - 7y \\ 8x + 5y \end{bmatrix} = \begin{bmatrix} 10 \\ 13 \end{bmatrix}$ $\begin{bmatrix} 2 & 4 \\ 4 & 5 \end{bmatrix} \begin{bmatrix} x \\ y \end{bmatrix} = \begin{bmatrix} 3 \\ 6 \end{bmatrix}$ $\begin{bmatrix} 5 & 3 & \vdots & 4 \\ 6 & 3 & \vdots & 7 \end{bmatrix}$

_____   _____   _____

3. $\begin{bmatrix} 3 & 6 & \vdots & 2 \\ 8 & -4 & \vdots & 8 \end{bmatrix}$ $\begin{bmatrix} 3 & 4 \\ 5 & 2 \end{bmatrix} \begin{bmatrix} x \\ y \end{bmatrix} = \begin{bmatrix} 9 \\ 4 \end{bmatrix}$ $\begin{bmatrix} 3 & 2 \\ 5 & 9 \end{bmatrix} \begin{bmatrix} x \\ y \end{bmatrix} = \begin{bmatrix} 1 \\ 5 \end{bmatrix}$

_____   _____   _____

4. $\begin{bmatrix} x - 6y \\ 3x + 4y \end{bmatrix} = \begin{bmatrix} 12 \\ 15 \end{bmatrix}$ $\begin{bmatrix} 1 & 2 & \vdots & 15 \\ 4 & 8 & \vdots & 6 \end{bmatrix}$ $\begin{bmatrix} 7x + 5y \\ 4x + 2y \end{bmatrix} = \begin{bmatrix} 5 \\ 3 \end{bmatrix}$

_____   _____   _____

#3 1. $\begin{bmatrix} 4 & 5 \\ 2 & -3 \end{bmatrix} \begin{bmatrix} 2 \\ 7 \end{bmatrix}$ $\begin{bmatrix} 2 & 3 \\ 3 & 1 \end{bmatrix} \begin{bmatrix} 1 \\ 8 \end{bmatrix}$ $\begin{bmatrix} 4 & 5 \\ 2 & -3 \end{bmatrix} \begin{bmatrix} 2 \\ 7 \end{bmatrix}$ $\begin{bmatrix} 2 & -3 \\ 4 & 5 \end{bmatrix} \begin{bmatrix} 7 \\ 2 \end{bmatrix}$
Decide how
it changed.   Multiply Add Multiply/Add Flip Switch   Multiply Add Multiply/Add Flip Switch

2. $\begin{bmatrix} 4 & 5 \\ 2 & -3 \end{bmatrix} \begin{bmatrix} 2 \\ 7 \end{bmatrix}$ $\begin{bmatrix} 4 & 5 \\ 6 & 3 \end{bmatrix} \begin{bmatrix} 2 \\ 9 \end{bmatrix}$ $\begin{bmatrix} 4 & 5 \\ 2 & -3 \end{bmatrix} \begin{bmatrix} 2 \\ 7 \end{bmatrix}$ $\begin{bmatrix} 8 & 10 \\ 2 & -3 \end{bmatrix} \begin{bmatrix} 4 \\ 7 \end{bmatrix}$

Multiply Add Multiply/Add Flip Switch   Multiply Add Multiply/Add Flip Switch

#4 Solve these with Cramer's Law.   Calculator?
                                                       yes   no

1. 5x + 2y = 3   1st step   ___ · ___ = ___
    x + 3y = 4   2nd step   ___ · ___ = ___   ▭ , ▭
                     last step ___ · ___ = ___

2. 2x + 2y = 8   1st step   ___ · ___ = ___
    x + 3y = 2   2nd step   ___ · ___ = ___   ▭ , ▭
                     last step ___ · ___ = ___

3. 6x + 3y = 1   1st step   ___ · ___ = ___
    x + 3y = 2   2nd step   ___ · ___ = ___   ▭ , ▭
                     last step ___ · ___ = ___

4. 8x + 2y = 5   1st step   ___ · ___ = ___
    x + 4y = 4   2nd step   ___ · ___ = ___   ▭ , ▭
                     last step ___ · ___ = ___

#5 Solve the 3 steps. What's the answer?   Calculator?
                                                       yes   no

1. What solves the 1st step with 3 variables? $4x + 2y + 5z = 10$

   Eliminate _____ $2x + 4y + z = 12$

          What happens next? $x + 3y + 5z = 14$

                                                 _____

   Eliminate _____

          What happens next?   _____

          What's the answer?   _____

                                                 _____

Ch 3 Ls 1, 2 How a Circle Moves   155

_____ #1 #2 ____/ 10 #3 #4 ____/ 8 R ____/ 5 T ____/ 23 _____
           Name   Checker

#1 1. What is the formula for a circle? _____

2. What do X and Y find in the circle formula? _____

3. Solve for 1 point of a circle. How many other points do you also know? _____

4. Find the point on this circle. 1st step? $2^2 + y^2 = 4^2$

Finish it. Estimate it to a 10th.   _____

Describe the graph.   _____

#2 1. How does it change for a different center? _____

2. What do H and K show? _____

3. How can you tell where a circle's center is? _____

4. Where is the center and radius? $(x + 1)^2 + (y - 2)^2 = 9$

Center is at ____, ____   Radius is ____

5. Where is the center and radius? $(x - 3)^2 + (y + 4)^2 = 25$

Center is at ____, ____   Radius is ____

6. Where is the center and radius? $(x + 4)^2 + (y - 6)^2 = 16$

Center is at ____, ____   Radius is ____

156

#3 Find the center and radius of each circle.   Calculator? yes no

1. $(x + 4)^2 + (y - 8)^2 = 9$ $(x - 4)^2 + (y - 3)^2 = 4$

   Center is at ___, ___   Radius is ___   Center is at ___, ___   Radius is ___

2. $(x + 5)^2 + (y - 3)^2 = 16$ $(x - 2)^2 + (y - 2)^2 = 5^2$

   Center is at ___, ___   Radius is ___   Center is at ___, ___   Radius is ___

#4 Find the point on this circle.   Calculator? yes no

1. 1st step? $-3^2 + y^2 = 6^2$ $x^2 + 5^2 = 4^2$

   Next?   _____   _____

   Square root?   _____   _____

   Describe
   the graph.   _____   _____

Center
_____

Radius
_____
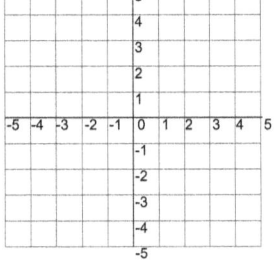

Center
_____

Radius
_____
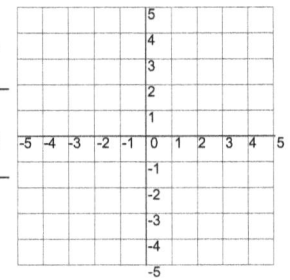

Review 1. What is the formula for a circle? _____

2. What do X and Y find in the circle formula? _____

3. Solve for 1 point of a circle. How many other points do you also know? _____

4. How does it change for a different center? _____

5. What do H and K show? _____

Ch 4 Ls 3 Is a point on a circle? 157

_____ #1 #2 ____/ 8 #3 ____/ 6 R ____/ 7 Total ____/ 21 _____
　　　Name　　　　　　　　　　　　　　　　　　　　　　　　　　　　　　　Checker

#1 1. How do you solve to find if a point is on a circle? _____

2. How can you tell if the point is on the circle? _____

3. Is 5, 4 on this circle? Solve it. $(x + 1)^2 + (y - 2)^2 = 9$

Put 5, 4 in the equation. Finish it. $(\_\_ + 1)^2 + (\_\_ - 2)^2 = 3^2$

Circle   Yes, it is.   No, it isn't.   _____

4. Is 1, 2 on this circle? Solve it. $(x + 2)^2 + (y + 3)^2 = 6^2$

Put 1, 2 in the equation. Finish it. $(\_\_ + 2)^2 + (\_\_ + 3)^2 = 6^2$

Circle   Yes, it is.   No, it isn't.   _____

#2 1. When you put a point in for a circle, what happens if it's correct? _____

_____

2. What circle is this? $(x - 4)^2 + (y + 1)^2 = 3^2$

　　Circle each.　　Left or Right ____　Up or Down ____　Radius ____

3. What circle is this? $(x + 5)^2 + (y - 3)^2 = 6^2$

　　Circle each.　　Left or Right ____　Up or Down ____　Radius ____

4. What circle is this? $(x - 3)^2 + (y - 1)^2 = 5^2$

　　Circle each.　　Left or Right ____　Up or Down ____　Radius ____

#3 Is the point on the circle?   Calculator?
                                        yes   no

1. (3, 4) $(x + 1)^2 + (y - 2)^2 = 9$ (4, 1) $(x + 1)^2 + (y - 2)^2 = 9$

   $(\underline{} + 1)^2 + (\underline{} - 2)^2 = 3^2$ $(\underline{} + 1)^2 + (\underline{} - 2)^2 = 3^2$

   Yes No _____   Yes No _____

2. (5, 6) $(x + 1)^2 + (y - 2)^2 = 9$ (-1, 3) $(x + 1)^2 + (y - 2)^2 = 9$

   $(\underline{} + 1)^2 + (\underline{} - 2)^2 = 3^2$ $(\underline{} + 1)^2 + (\underline{} - 2)^2 = 3^2$

   Yes No _____   Yes No _____

3. (-2, 5) $(x + 1)^2 + (y - 2)^2 = 9$ (7, -2) $(x + 1)^2 + (y - 2)^2 = 9$

   $(\underline{} + 1)^2 + (\underline{} - 2)^2 = 3^2$ $(\underline{} + 1)^2 + (\underline{} - 2)^2 = 3^2$

   Yes No _____   Yes No _____

Review 1. How do you solve to find if a point is on a circle? _____   Calculator?
        2. How can you tell if the point is on the circle? _____   yes   no
        3. What rule reminds you how an equation shows a circle moves? _____

4. (1, Y) $(1 + 1)^2 + (y - 2)^2 = 3^2$ (X, 3) $(x + 1)^2 + (3 - 2)^2 = 5^2$

Find what   _____   _____
point it is.
                _____   _____

5. (3, Y) $(3 + 2)^2 + (y + 3)^2 = 7^2$ (X, 4) $(x + 5)^2 + (4 - 1)^2 = 9^2$

                _____   _____

                _____   _____

Ch 4 Ls 4 Distance Between Center and a Point 159

_____ #1 #2 ____/ 7 #3 ____/ 3 R ____/ 4 Total ____/ 14 _____
Name   Checker

#1 1. What does the Distance Formula do? _____
2. Name 3 steps to Distance Formula. _____
3. What is the distance formula? _____
4. How is distance formula different from circle formula? _____

#2 1. What's the 1st step to find distance?   (0, 2)   (5, 4)

    What's 1st in SSS?   Xs ___ - ___ = ___   Ys ___ - ___ = ___

    What next in SSS?   X squared is ___   Y squared is ___

    What last in SSS?   ___ + ___ = ___

    Square root of ___ is about ___

2. What's the 1st step to find distance?   (3, 3)   (7, 1)

    What's 1st in SSS?   Xs ___ - ___ = ___   Ys ___ - ___ = ___

    What next in SSS?   X squared is ___   Y squared is ___

    What last in SSS?   ___ + ___ = ___

    Square root of ___ is about ___

                                        Center   Point.

3. Solve the formula. Xs and Ys?   (1, 2) ——— (6, 4)

Solve inside each parentheses. $(\_\_ - \_\_)^2 + (\_\_ - \_\_)^2 = r^2$

Solve the squared numbers. Estimate it. $(\_\_)^2 + (\_\_)^2 = r^2$

                                             _____
                                             _____

#3   Find the radius using the distance formula.   Calculator?
                                                                    yes   no

                                    Center   Point
1. What is the 2nd and 1st X and Y? (2, -1) —— (3, 3)

   Solve inside each parentheses. $(\_ - \_)^2 + (\_ - \_)^2 = r^2$

   Solve the squared numbers. Estimate it. $(\_)^2 + (\_)^2 = r^2$

                                              _____ = r

                                    Center   Point
2. What is the 2nd and 1st X and Y? (-3, 2) —— (4, 2)

   Solve inside each parentheses. $(\_ - \_)^2 + (\_ - \_)^2 = r^2$

   Solve the squared numbers. Estimate it. $(\_)^2 + (\_)^2 = r^2$

                                              _____ = r

                                    Center   Point
3. What is the 2nd and 1st X and Y? (-2, 2) —— (7, 1)

   Solve inside each parentheses. $(\_ - \_)^2 + (\_ - \_)^2 = r^2$

   Solve the squared numbers. Estimate it. $(\_)^2 + (\_)^2 = r^2$

                                              _____ = r

Review 1. What does the Distance Formula do? _____

        2. Name 3 steps to Distance Formula. _____

        3. What is the distance formula? _____

        4. How is distance formula different from circle formula? _____

Review Problems   161

_____   #1 - #3 ____ / 9   Back ____ / 7   Total ____ / 16
Name

1. Circle _____

2. Think Opposite _____

3. Distance Formula _____

_____

#2 Find the center and radius of each circle.

1. $(x + 1)^2 + (y - 2)^2 = 9$ $(x - 4)^2 + (y - 3)^2 = 4$

   Center is at ___, ___ Radius is ___   Center is at ___, ___ Radius is ___

2. $(x + 5)^2 + (y - 3)^2 = 16$ $(x - 2)^2 + (y - 2)^2 = 5^2$

   Center is at ___, ___ Radius is ___   Center is at ___, ___ Radius is ___

#3 If Y is 2, what would X be?

1. 1st step? $(3 + 2)^2 + (y - 1)^2 = 3^2$ $(1 + 3)^2 + (y - 3)^2 = 2^2$

   Next?   _____   _____

Square root?   _____   _____

Describe
the graph.   _____   _____

Center
_____
Radius
_____
Point
_____

#4 Is the point on the circle?   Calculator?
                                                                                          yes   no

1. (5, 6) $(x + 2)^2 + (y + 3)^2 = 9$ (6, 1) $(x + 4)^2 + (y - 3)^2 = 25$

    $(\_\_ + 2)^2 + (\_\_ - 3)^2 = 3^2$ $(\_\_ + 4)^2 + (\_\_ - 3)^2 = 5^2$

    Yes No _____   Yes No _____

2. (7, 4) $(x + 9)^2 + (y + 7)^2 = 4$ (-3, -4) $(x + 3)^2 + (y - 4)^2 = 16$

    $(\_\_ + 9)^2 + (\_\_ + 7)^2 = 2^2$ $(\_\_ + 3)^2 + (\_\_ - 4)^2 = 4^2$

    Yes No _____   Yes No _____

#5 Find the radius using the distance formula.   Calculator?
                                                                                          yes   no
                                          Center   Point.
1. What is the 2nd and 1st X and Y?   (2, 4) —— (5, 2)

    Solve inside each parentheses. $(\_\_ - \_\_)^2 + (\_\_ - \_\_)^2 = r^2$

    Solve the squared numbers. Estimate it. $(\_\_)^2 + (\_\_)^2 = r^2$

                                                     _____ = r

                                          Center   Point.
2. What is the 2nd and 1st X and Y?   (-4, 1) —— (8, 3)

    Solve inside each parentheses. $(\_\_ - \_\_)^2 + (\_\_ - \_\_)^2 = r^2$

    Solve the squared numbers. Estimate it. $(\_\_)^2 + (\_\_)^2 = r^2$

                                                     _____ = r

                                          Center   Point.
3. What is the 2nd and 1st X and Y?   (3, -7) —— (6, 4)

    Solve inside each parentheses. $(\_\_ - \_\_)^2 + (\_\_ - \_\_)^2 = r^2$

    Solve the squared numbers. Estimate it. $(\_\_)^2 + (\_\_)^2 = r^2$

                                                     _____ = r

Ch 4 Ls 1 Circles and Intercepts   163

_____ #1 #2 ____ / 7 #3 ____ / 4 R ____ / 3 Total ____ / 14 _____
Name   Checker

#1 1. If an equation equals a number, what happens to find the radius? _____
_____

2. How many intercepts can a circle have? _____

3. What do you use to find a X intercept? _____

#2 1. How do you find the X intercept for this circle? $(x - 3)^2 + (y - 1)^2 = 4$

Put 0 in for ___. First step? $(x - 3)^2 + (\_\_ - 1)^2 = 4$

What can you square?   _____

Solve the next step. What is left?   _____

How do you solve a squared binomial?   _____

_____

2. How do you find the Y intercept for this circle? $(x - 4)^2 + (y - 0)^2 = 4$

Put 0 in for ___. First step? $(\_\_ - 4)^2 + (y - 0)^2 = 4$

What can you square?   _____

Solve the next step. What is left?   _____

How do you solve a squared binomial?   _____

_____

#3 Find the X intercepts.   Calculator? yes   no

1. $(x - 5)^2 + (y - 1)^2 = 7$ $(x - 4)^2 + (y - 4)^2 = 6$

   $(x - 5)^2 + (\underline{} - 1)^2 = 7$ $(x - 4)^2 + (\underline{} - 4)^2 = 6$

   _____   _____

   _____   _____

   _____   _____

   _____   _____

4. $(x - 1)^2 + (y - 2)^2 = 3$ $(x - 3)^2 + (y - 1)^2 = 8$

How many Y intercepts are there.

   $(\underline{} - 1)^2 + (y - 2)^2 = 3$ $(\underline{} - 3)^2 + (y - 1)^2 = 8$

   _____   _____

   _____   _____

   _____   _____

   _____   _____

Review 1. If an equation equals a number, what happens to find the radius? _____

_____

2. How many intercepts can a circle have? _____

3. What do you use to find a X intercept? _____

Ch 4 Ls 2 Circle and Complete the Sqaure   165

_____ #1 #2 ____/ 8 #3 ____/ 4 R ____/ 7 Total ____/ 19 _____
        Name   Checker

Example $x^2 - 4x + y^2 - 2y = 0$

#1 1. What does x and y squared need to be a circle equation? _____

2. Name 2 steps to Complete the Square with each term. _____
_____

3. What step changes an equation to perfect binomials? _____

4. What happens to what's added? _____

#2 1. Complete the square for X. $x^2 - 4x + (y - 2)^2 = 0$

Make a perfect binomial.
What adds to the right? $x^2 - 4x + \underline{}$

$(x \quad)^2 + (y - 2)^2 = \underline{}$

2. Complete the square for X. $x^2 - x + (y + 1)^2 = 0$

Make a perfect binomial.
What adds to the right? $x^2 - x + \underline{}$

$(x \quad)^2 + (y + 1)^2 = \underline{}$

3. Complete the square for Y. $(x + 1)^2 + y^2 + 2y = 0$

Make a perfect binomial.
What adds to the right? $y^2 + 2y + \underline{}$

$(x + 1)^2 + (y \quad)^2 = \underline{}$

4. Complete the square for Y. $(x + 4)^2 + y^2 + 3y = 0$

Make a perfect binomial.
What adds to the right? $y^2 + 3y + \underline{}$

$(x + 4)^2 + (y \quad)^2 = \underline{}$

#3 1. $x^2 - 4x + (y-2)^2 = 0$ $\qquad$ $(x^2 - 6) + y^2 - 8y = 0$ $\qquad$ Calculator?
   $x^2 - 4x + \underline{}$ $\qquad\qquad\qquad$ $y^2 - 8y + \underline{}$ $\qquad\qquad$ yes   no

   $x^2 - 4x \underline{} + (y-2)^2 = \underline{}$ $\qquad$ $(x-6)^2 + y^2 - 2y \underline{} = \underline{}$

   $(x)^2 + (y-2)^2 = \underline{}$ $\qquad\qquad$ $(x-6)^2 + (y)^2 = \underline{}$

2. $x^2 - 1x + (y-6)^2 = 0$ $\qquad\qquad$ $(x^2 - 4) + y^2 - 2y = 0$

   $x^2 - 1x + \underline{}$ $\qquad\qquad\qquad$ $y^2 - 2y + \underline{}$

   $x^2 - 1x \underline{} + (y-6)^2 = \underline{}$ $\qquad$ $(x-4)^2 + y^2 - 2y \underline{} = \underline{}$

   $(x)^2 + (y-6)^2 = \underline{}$ $\qquad\qquad$ $(x-4)^2 + (y)^2 = \underline{}$

Review 1. What does x and y squared need to be a circle equation? _____ Calculator?
2. What step changes an equation to perfect binomials? _____ yes   no
3. Name 2 steps to Complete the Square with each term. _____

4. $x^2 - 8x + (y-2)^2 = 0$ $\qquad\qquad$ $(x^2 - 4) + y^2 - 10y = 0$

   $x^2 - 8x + \underline{}$ $\qquad\qquad\qquad$ $y^2 - 10y + \underline{}$

   $x^2 - 8x \underline{} + (y-2)^2 = \underline{}$ $\qquad$ $(x-4)^2 + y^2 - 10y \underline{} = \underline{}$

   $(x-8)^2 + (y)^2 = \underline{}$ $\qquad\qquad$ $(x-4)^2 + (y)^2 = \underline{}$

5. $x^2 - 5x + (y-3)^2 = 0$ $\qquad\qquad$ $(x^2 - 8) + y^2 - 12y = 0$

   $x^2 - 5x + \underline{}$ $\qquad\qquad\qquad$ $y^2 - 12y + \underline{}$

   $x^2 - 5x \underline{} + (y-3)^2 = \underline{}$ $\qquad$ $(x-8)^2 + y^2 - 12y \underline{} = \underline{}$

   $(x-5)^2 + (y)^2 = \underline{}$ $\qquad\qquad$ $(x-8)^2 + (y)^2 = \underline{}$

_____ #1 #2 ____ / 9 #3 ____ / 3 R ____ / 4 Total ____ / 16 _____
               Name   Checker

#1 1. How do you find the midpoint of 2 points? _____

2. Name 3 steps to an equation from 2 endpoints. _____

_____

3. Find the Xs and the Ys. What's the midpoint? $(-8, 2)$ $(2, 5)$

___ ÷ 2 = ___   ___ ÷ 2 = ___   Midpoint: _____

4. Altogether, what's the midpoint? $(-6, 1)$ $(3, 0)$

___ ÷ 2 = ___   ___ ÷ 2 = ___   Midpoint: _____

5. Altogether, what's the midpoint? $(-2, 4)$ $(9, -1)$

___ ÷ 2 = ___   ___ ÷ 2 = ___   Midpoint: _____

#2 1. What is the 1st step? **Write an equation for a circle if it has end points for it's diameter at $(-2, 3)$ and $(2, -5)$.**

What is the radius? ___ ÷ 2 = ___   ___ ÷ 2 = ___   Center is at: _____

Make an equation.   ___ ÷ 2 = ___   ___ ÷ 2 = ___   Center is at: _____

_____

2. What is the 1st step? **Write an equation for a circle if it has end points for it's diameter at $(-4, 1)$ and $(2, -7)$.**

What is the radius? ___ ÷ 2 = ___   ___ ÷ 2 = ___   Center is at: _____

Make an equation.   ___ ÷ 2 = ___   ___ ÷ 2 = ___   Center is at: _____

_____

#3 Find the midpoint from the 2 points.   Calculator? yes no

1. Find the Xs and the Ys. What's the midpoint?   (1, -4) (6, 5)

   ___ ÷ 2 = ___   ___ ÷ 2 = ___   Midpoint: _____

2. What's the midpoint?   (-3, 1) (6, 2)

   ___ ÷ 2 = ___   ___ ÷ 2 = ___   Midpoint: _____

3. What's the midpoint?   (-7, 3) (3, 4)

   ___ ÷ 2 = ___   ___ ÷ 2 = ___   Midpoint: _____

Review 1. How do you find the midpoint of 2 points? _____   Calculator? yes no

2. Name 3 steps to an equation from 2 endpoints. _____

_____

3. What is the 1st step?   **Write an equation for a circle if it has end points for it's diameter at (1, 5) and (3, -5).**

What is the radius?   ___ ÷ 2 = ___   ___ ÷ 2 = ___   Center is at: _____

Make an equation.   ___ ÷ 2 = ___   ___ ÷ 2 = ___   Center is at: _____

_____

4. What is the 1st step?   **Write an equation for a circle if it has end points for it's diameter at (-4, 3) and (8, -6).**

What is the radius?   ___ ÷ 2 = ___   ___ ÷ 2 = ___   Center is at: _____

Make an equation.   ___ ÷ 2 = ___   ___ ÷ 2 = ___   Center is at: _____

_____

Ch 4 Ls 4 Make an equation from a graph.   169

_____ #1 #2 ___ / 8 #3 ___ / 4 R ___ / 6 Total ___ / 20 _____
       Name   Checker

#1 1. How do you solve if a point is on a circle? _____

2. How can you tell if the point is on the circle? _____

#2 1.

Make an equation for each circle.

#3

Make an equation
for these circles.

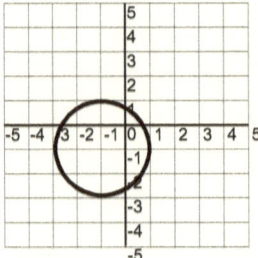
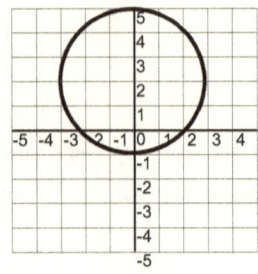

Calculator?
yes   no

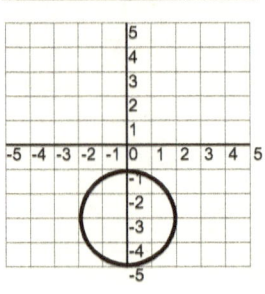
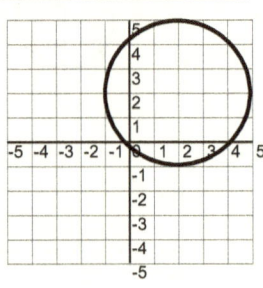

Review 1. How do you solve to find if a point is on a circle? _____

2. How can you tell if the point is on the circle? _____

Story Problems

3. **A water sprinkler can spray water at a maximum distance of 15 m in all directions. What area of the lawn can this sprinkler irriguate?**   _____ sq m

4. A circular pizza costs Rs 300. What is the cost of 1 square centimeter if the diameter of the pizza is 40 cm?   Rs _____

5. How much fencing is needed for a circular flower garden that has an area of 6 square meters? (round to nearest meter.)   _____ m

6. A circular table has a dimeter of 400 cm. A circular tablecloth hangs over the table 8 cm around the table. What is the area of the tablecloth?   _____ sq m

Review Problems 171

_____ #1 #2 #3 _____ / 10 #4 #5 _____ / 7 Total _____ / 17
Name

1. Y intercept (circle) _____

2. X intercept (circle) _____

3. Midpoint _____

#2 1. $(x-5)^2 + (y-1)^2 = 7$ $(x-4)^2 + (y-4)^2 = 6$

Find the X
intercepts. $(x-5)^2 + (\_\_-1)^2 = 7$ $(x-4)^2 + (\_\_-4)^2 = 6$

_____   _____
_____   _____
_____   _____
_____   _____

#3 1. $x^2 - 4x + (y-2)^2 = 0$ $(x^2 - 6) + y^2 - 8y = 0$

Solve with
complete   _____   _____
the square.   _____   _____
              _____   _____

2. $x^2 - 1x + (y-6)^2 = 0$ $(x^2 - 4) + y^2 - 2y = 0$

_____   _____
_____   _____
_____   _____

#4 Find the midpoint from the 2 points.   Calculator? yes no

1. Find the Xs and the Ys. What's the midpoint?   (1, -4) (6, 5)

   ___ ÷ 2 = ___   ___ ÷ 2 = ___   Midpoint: _____

2. What's the midpoint?   (-3, 1) (6, 2)

   ___ ÷ 2 = ___   ___ ÷ 2 = ___   Midpoint: _____

3. What's the midpoint?   (-7, 3) (3, 4)

   ___ ÷ 2 = ___   ___ ÷ 2 = ___   Midpoint: _____

#5 If you know the answer, you still need to work it out.   Calculator? yes no

1. Is -1, -1 on this circle? Solve it. $(x - 1)^2 + (y - 4)^2 = 3^2$

   Finish it. $(\_\_ - 1)^2 + (\_\_ - 4)^2 = 3^2$

   Circle   **Yes, it is. No, it isn't.**   _____

2. Is 4, 6 on this circle? Solve it. $(x + 2)^2 + (y + 3)^2 = 5$

   Put 4, 6 in the equation. Finish it. $(\_\_ + 2)^2 + (\_\_ + 3)^2 = 5^2$

   Circle   **Yes, it is. No, it isn't.**   _____

3. Is 0, 1 on this circle? Solve it. $(x - 3)^2 + (y + 4)^2 = 3^2$

   Put 0, 1 in the equation. Finish it. $(\_\_ - 3)^2 + (\_\_ + 4)^2 = 3^2$

   Circle   **Yes, it is. No, it isn't.**   _____

4. Is 3, 7 on this circle? Solve it. $(x + 2)^2 + (y + 3)^2 = 5$

   Put 3, 7 in the equation. Finish it. $(\_\_ + 2)^2 + (\_\_ + 3)^2 = 5^2$

   Circle   **Yes, it is. No, it isn't.**   _____

Ch 5 Ls 1 How ellipse is different from circle. 173

_____ #1 #2 ____/ 7 #3 ____/ 3 R ____/ 5 Total ____/ 15 _____
         Name   Checker

#1 1. How long and wide is an ellipse? _____
    2. How do 2 foci points find each point on the ellipse? _____
    3. How does d and e find the distance of the ellipse? _____
    4. What do D and E add upto? _____

   #2 1. Measure this ellipse.
          How long and wide is the ellipse?

        A is ___, so 2A is ___.   B is ___, so 2B is ___.

        2. Measure this ellipse.
        How long and wide is the ellipse?

        A is ___, so 2A is ___.   B is ___, so 2B is ___.

        3. Measure this ellipse.
        How long and wide is the ellipse?

        A is ___, so 2A is ___.   B is ___, so 2B is ___.

174

#3 1. How long and wide is the ellipse?   Calculator?
                                                                            yes   no

Measure this ellipse.   A is ___, so 2A is ___.
Draw the foci in.
                          B is ___, so 2B is ___.

2. Measure this ellipse. Draw the foci in.
   How long and wide is the ellipse?

                          A is ___, so 2A is ___.

                          B is ___, so 2B is ___.

3. Measure this ellipse. Draw the foci in.
   How long and wide is the ellipse?

                          A is ___, so 2A is ___.

                          B is ___, so 2B is ___.

Review 1. How long and wide is an ellipse? _____   Calculator?
                                                                                yes   no
        2. How do 2 foci points find each point on the ellipse? _____

        3. How does d and e find the distance of the ellipse? _____

        4. What do D and E add up to? _____

        5. Draw an ellipse. Decide A and B.

            A is ___ and B is ___.

                A is ___, so 2A is ___.

                B is ___, so 2B is ___.

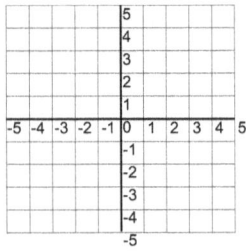

Ch 5 Ls 2 How to find each point on an ellipse. 175

_____ #1 #2 ____/ 8 #3 ____/ 2 R ____/ 7 Total ____/ 17 _____
　　　　　Name　　　　　　　　　　　　　　　　　　　　　　　　　　　　　　　　　　Checker

#1 1. What formula measures each point from each focus? _____

　　　2. Name 2 steps to draw a focus in. _____

　　　3. What does C measure? _____

　　　4.　First, draw both Foci in.
　　　　　You know A. Draw B and C.

　　　　　How long are they?
　　　　　Make the equation below.

　　　　　How long is C?

____ + ____ = ____　　　____ + ____ = ____　　　C = ____

#2 1. What letter finds the center to the foci point? _____

　　　2. What formula does this triangle use? _____

　　　3.　Draw both Foci in.
　　　　　You know A. Draw B and C.

　　　　　How long are they?
　　　　　Make the equation below.

　　　　　How long is C?

____ + ____ = ____　　　____ + ____ = ____　　　C = ____

　　　4.　Draw both Foci in.
　　　　　You know A. Draw B and C.

　　　　　How long are they?
　　　　　Make the equation below.

　　　　　How long is C?

____ + ____ = ____　　　____ + ____ = ____　　　C = ____

#3 Draw the foci in. Find what C is.   Calculator?
                                                      yes   no

1.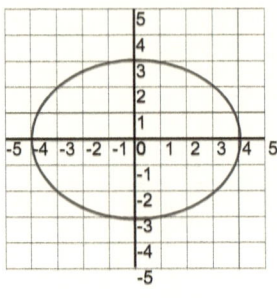

   ___ + ___ = ___   ___ + ___ = ___
   ___ + ___ = ___   ___ + ___ = ___
       C = ___   C = ___

Review 1. What formula measures each point from each focus? _____   Calculator?
                                                                                    yes   no
2. Name 2 steps to draw a focus in. _____

3. What does C measure? _____

4. What letter finds the center to the foci point? _____

5. What formula does this triangle use? _____

6.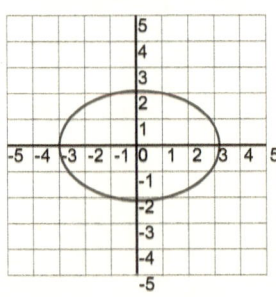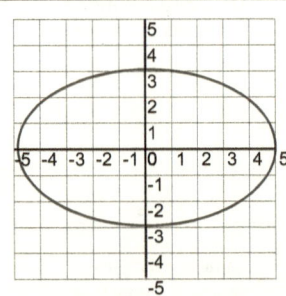

   ___ + ___ = ___   ___ + ___ = ___
   ___ + ___ = ___   ___ + ___ = ___
       C = ___   C = ___

Ch 5 Ls 3 How to find foci.   177

_____ #1 #2 ____ / 8   #3 ____ / 4   R ___ / 12   Total ____ / 24 _____
           Name   Checker

#1 1. What numbers is the eccentricity between? _____

2. What happens to the focii as the ellipse gets skinnier? _____

3. What happens to the focii when A and B are the same? _____

4. What's the formula for eccentricity? _____

#2 1. Find the eccentricity. If A is 6 and B is 3, how far is the focus from the center point?

3, 6, C (right triangle)

Solve the squared numbers.   ___^{2} = ___^{2} − ___^{2}

Subtract the terms.   ___^{2} = ___ − ___

Estimate how long C is.   ___^{2} = ___

What is the eccentricity?   ___ = ___

___ ÷ ___ = ___

2. What if you already know C? Mentally find the eccentricity.   **C is 5, A is 10**

___ ÷ ___ = ___

3. What is the eccentricity?   **C is 6, A is 8**

___ ÷ ___ = ___

4. What is the eccentricity?   **C is 3, A is 6**

___ ÷ ___ = ___

#3 Estimate how long C is.

Calculator? yes no

1. You know A and B. How far is the focus from the center point?

Estimate C

Triangle: legs 3 and 6, hypotenuse C

___² = ___² - ___²

___² = ___ - ___

___² = ___

___ = ___

Triangle: legs 4 and 8, hypotenuse C

___² = ___² - ___²

___² = ___ - ___

___² = ___

___ = ___

2.

Estimate C

Triangle: legs 2 and 5, hypotenuse C

___² = ___² - ___²

___² = ___ - ___

___² = ___

___ = ___

Triangle: legs 6 and 9, hypotenuse C

___² = ___² - ___²

___² = ___ - ___

___² = ___

___ = ___

Review

1. What letter finds the center to the foci point? _____

Calculator? yes no

2. What formula does this triangle use? _____

3. What is eccentricity? _____

4. What happens to the focii as the ellipse gets skinnier? _____

5. What happens to the focii when A and B are the same? _____

6. What's the formula for eccentricity? _____

Find the eccentricity.

7. C is 5, A is 10   C is 1, A is 4   C is 3, A is 8

___ ÷ ___ = ___   ___ ÷ ___ = ___   ___ ÷ ___ = ___

8. C is 2, A is 7   C is 3, A is 9   C is 2, A is 4

___ ÷ ___ = ___   ___ ÷ ___ = ___   ___ ÷ ___ = ___

Ch 5 Ls 4 The Ellipse Formula   179

_____ #1 #2 ____/ 9 #3 ____/ 7 R ____/ 5 Total ____/ 12 _____
             Name   Checker

#1 1. What is the left side of an ellipse formula? _____
    2. What does the ellipse equation equal? _____
    3. If an ellipse equation doesn't equal 1, how do you solve it? _____
    4. If B is bigger than A, what does the ellipse look like? _____
    5. Name 2 things the left side has to have for an ellipse. _____

    _____

#2 1. What are A and B? Describe this ellipse. $\dfrac{x^2}{2^2} + \dfrac{y^2}{1^2} = 1$

         A is ____. B is ____. The ellipse is _____.

    2. What are A and B? Describe this ellipse. $\dfrac{x^2}{4} + \dfrac{y^2}{9} = 1$

         A is ____. B is ____. The ellipse is _____.

    3. What are A and B? Describe this ellipse. $\dfrac{x^2}{16} + \dfrac{y^2}{9} = 1$

         A is ____. B is ____. The ellipse is _____.

    4. What's the 1st step to get a standard equation? $2x^2 + 4y^2 = 8$

       Divide to get ____ on the right. What's the next step? $\dfrac{2x^2}{8} + \dfrac{4y^2}{8} = 1$

       Divide to get _____. What's next? $\dfrac{x^2}{4} + \dfrac{y^2}{2} = 1$

       Square Roots in _____. $\dfrac{x^2}{2^2} + \dfrac{y^2}{1.4^2} = 1$

#3 Change to a standard equation, then describe each ellipse.   Calculator? yes   no

1. What's the 1st step to get a standard equation?

   $2x^2 + 8y^2 = 16$

   Divide to get ___ on the right. What's the next step?

   $\dfrac{2x^2}{16} + \dfrac{8y^2}{16} = 1$

   Divide to get _____. What's next?

   $\dfrac{x^2}{8} + \dfrac{y^2}{4} = 1$

   Square Roots in each _____.

   $\dfrac{x^2}{2.8^2} + \dfrac{y^2}{2^2} = 1$

2. $\dfrac{x^2}{2^2} + \dfrac{y^2}{6^2} = 1$ $\dfrac{x^2}{5^2} + \dfrac{y^2}{3^2} = 1$

   Ellipse is ___ long and ___ wide.   Ellipse is ___ long and ___ wide.

3. $\dfrac{x^2}{9^2} + \dfrac{y^2}{3^2} = 1$ $\dfrac{x^2}{7^2} + \dfrac{y^2}{2^2} = 1$

   Ellipse is ___ long and ___ wide.   Ellipse is ___ long and ___ wide.

4. $\dfrac{x^2}{7^2} + \dfrac{y^2}{5^2} = 1$ $\dfrac{x^2}{8^2} + \dfrac{y^2}{4^2} = 1$

   Ellipse is ___ long and ___ wide.   Ellipse is ___ long and ___ wide.

Review 1. What is the left side of an ellipse formula? _____

2. What does the ellipse equation equal? _____

3. If an ellipse equation doesn't equal 1, how do you solve it? _____

4. If B is bigger than A, what does the ellipse look like? _____

5. Name 2 things the left side has to have for an ellipse. _____

_____

Ch 5 Ls 5 Complete the square with ellipses.   181

_____ #1 #2 ___/7 #3 ___/4 R ___/5 Total ___/16 _____
Name   Checker

Example $x^2 - 4x + y^2 - 2y = 0$

#1 1. How did circle change for a different center? _____
2. What is the ellipse equation when it moves from 0? _____
3. Name 2 steps so ellipse equation has perfect binomials? _____
4. How is ellipse different from circle equation? _____
5. What do ellipses have to equal? _____

#2 1. Complete the square for X. $x^2 - 4x + \dfrac{(y-1)^2}{10^2} = 0$

Make a perfect binomial.
What adds to the right? $x^2 - 4x + \underline{} + \dfrac{(y-1)^2}{10^2} = \underline{}$

$\dfrac{(x)^2}{\underline{}} + \dfrac{(y-1)^2}{10^2} = \underline{}$

$\dfrac{(x)^2}{\underline{}} + \dfrac{(y-1)^2}{10^2} = \underline{}$

2. Complete the square for X. $x^2 - 6x + \dfrac{(y-1)^2}{4^2} = 0$

Make a perfect binomial.
What adds to the right? $x^2 - 6x + \underline{} + \dfrac{(y-1)^2}{4^2} = \underline{}$

$\dfrac{(x)^2}{\underline{}} + \dfrac{(y-1)^2}{4^2} = \underline{}$

$\dfrac{(x)^2}{\underline{}} + \dfrac{(y-1)^2}{4^2} = \underline{}$

#3 Complete the Square.

1. $\dfrac{(y-1)^2}{10} + y^2 + 8y = 0$ $x^2 - 5x - 2 + \dfrac{(y-2)^2}{12} = 0$ Calculator? yes no

    $\dfrac{(y-1)^2}{10} + y^2 + 8y \;\underline{\quad} = \underline{\quad}$ $x^2 - 5x + \underline{\quad} + \dfrac{(y-2)^2}{12} = \underline{\quad}$

    $\dfrac{(y-1)^2}{10} + \dfrac{(y\;\;\;)^2}{\underline{\quad}} = \underline{\quad}$ $\dfrac{(x\;\;\;)^2}{\underline{\quad}} + \dfrac{(y-2)^2}{12} = \underline{\quad}$

    $\dfrac{(y-1)^2}{10} + \dfrac{(y\;\;\;)^2}{\underline{\quad}} = \underline{\quad}$ $\dfrac{(x\;\;\;)^2}{\underline{\quad}} + \dfrac{(y-2)^2}{12} = \underline{\quad}$

2. $\dfrac{(y-1)^2}{8} + y^2 + 9y = 0$ $x^2 - 2x - 1 + \dfrac{(y-3)^2}{20} = 0$

    $\dfrac{(y-1)^2}{8} + y^2 + 9y \;\underline{\quad} = \underline{\quad}$ $x^2 - 2x + \underline{\quad} + \dfrac{(y-3)^2}{20} = \underline{\quad}$

    $\dfrac{(y-1)^2}{8} + \dfrac{(y\;\;\;)^2}{\underline{\quad}} = \underline{\quad}$ $\dfrac{(x\;\;\;)^2}{\underline{\quad}} + \dfrac{(y-3)^2}{20} = \underline{\quad}$

    $\dfrac{(y-1)^2}{8} + \dfrac{(y\;\;\;)^2}{\underline{\quad}} = \underline{\quad}$ $\dfrac{(x\;\;\;)^2}{\underline{\quad}} + \dfrac{(y-3)^2}{20} = \underline{\quad}$

Review 1. How did circle change for a different center? _____   Calculator? yes no

2. What is the ellipse equation when it moves from 0? _____

3. Name 2 steps so ellipse equation has perfect binomials? _____

4. How is ellipse different from circle equation? _____

5. What do ellipses have to equal? _____

Ch 5 Ls 6 Area and Perimeter of an Ellipse   183

_____ #1 #2 ____ / 7 #3 ____ / 5 R ____ / 3 Total ____ / 15 _____
         Name   Checker

#1 1. What's the formula for the area of an ellipse? _____

2. What's the formula for the perimeter of an ellipse? _____

3. Multiply 2 Pi for perimeter. What is 2 pi? _____

#2 1. Find the area on this ellipse. 1st step? $2^2 + y^2 = 4^2$

Finish it. Estimate it to a 10th.   _____

_____

2. Find the area on this ellipse 1st step? $3^2 + y^2 = 6^2$

Finish it. Estimate it to a 10th.   _____

_____

3. Find the perimeter for this ellipse. 1st step? $1^2 + y^2 = 3^2$

Finish it. Estimate it to a 10th. $6.3 \sqrt{\dfrac{2^2 + 2^2}{2}}$ _____

_____

4. Find the perimeter on this ellipse. 1st step? $5^2 + y^2 = 8^2$

Finish it. Estimate it to a 10th. $6.3 \sqrt{\dfrac{2^2 + 2^2}{2}}$ _____

_____

#3 1. Find the area of this ellipse. 1st step? $4^2 + y^2 = 6^2$ Calculator? yes no

Find the area or perimeter.   Finish it. Estimate it to a 10th.   _____

_____

2. Find the area of this ellipse. 1st step? $3^2 + y^2 = 10^2$

Finish it. Estimate it to a 10th.   _____

_____

3. Find the area of this ellipse. 1st step? $5^2 + y^2 = 9^2$

Finish it. Estimate it to a 10th.   _____

_____

4. Find the perimeter on this ellipse. 1st step? $4^2 + y^2 = 8^2$

Finish it. Estimate it to a 10th.   _____

_____

5. Find the perimeter on this ellipse. 1st step? $5^2 + y^2 = 2^2$

Finish it. Estimate it to a 10th.   _____

_____

Review 1. What's the formula for the area of an ellipse? _____   Calculator?

2. What's the formula for the perimeter of an ellipse? _____   yes no

3. Multiply 2 Pi. What is 2 pi? _____

Review Problems 185

_____ #1 - #4 ____/ 14   #4 - #7 ____/ 12   Total ____/ 26
                Name

1. Ellipse _____

2. Foci _____

3. Ellipse Equation _____

4. Ellipse Area _____

5. Ellipse Perimeter _____

#2 Draw the foci in. How long and wide is the ellipse?

1.

2.

3.

A is ___, so 2A is ___.   A is ___, so 2A is ___.   A is ___, so 2A is ___.

B is ___, so 2B is ___.   B is ___, so 2B is ___.   B is ___, so 2B is ___.

#3 Draw the foci in. Estimate B when A is 2 and 4.

1.

2.

3.

When A is 1.5, B is _____.   When A is 4, B is _____.   When A is 2, B is _____.

When A is -1, B is _____.   When A is -1, B is _____.   When A is -3, B is _____.

#4 Find the eccentricity.

1. **C is 5, A is 10**   2. **C is 1, A is 4**   3. **C is 3, A is 8**

   ___ ÷ ___ = ___   ___ ÷ ___ = ___   ___ ÷ ___ = ___

Continued from part 4.

Calculator? yes no

2. C is 2, A is 7   C is 3, A is 9   C is 2, A is 4

___ ÷ ___ = ___   ___ ÷ ___ = ___   ___ ÷ ___ = ___

#5 How long and wide is each ellipse?

1. $\dfrac{x^2}{2^2} + \dfrac{y^2}{6^2} = 1$ It's ___ long and ___ wide.

   $\dfrac{x^2}{5^2} + \dfrac{y^2}{3^2} = 1$ It's ___ long and ___ wide.

2. $\dfrac{x^2}{9^2} + \dfrac{y^2}{3^2} = 1$ It's ___ long and ___ wide.

   $\dfrac{x^2}{7^2} + \dfrac{y^2}{2^2} = 1$ It's ___ long and ___ wide.

#6 Complete the Square to solve these.

1. $\dfrac{(y-1)^2}{10} + y^2 + 6y = 0$

   $\dfrac{(y-1)^2}{10} + y^2 + 6y \ \_\_\_ = \_\_\_$

   $\dfrac{(y-1)^2}{10} + \dfrac{(y\ \_\_\_)^2}{\ \ } = \_\_\_$

   $\dfrac{(y-1)^2}{10} + \dfrac{(y\ \_\_\_)^2}{\ \ } = \_\_\_$

   $x^2 - 4x - 1 + \dfrac{(y-1)^2}{10} = 0$

   $x^2 - 4x + \_\_\_ + \dfrac{(y-1)^2}{10} = \_\_\_$

   $(x\ \_\_\_)^2 + \dfrac{(y-1)^2}{10} = \_\_\_$

   $(x\ \_\_\_)^2 + \dfrac{(y-1)^2}{10} = \_\_\_$

#7 Find the area or perimeter of each ellipse.

1. According to science, planets have elliptical orbits, with the sun at one of the foci. Pluto is 7.4 billion kilometers from the sun. The closest it gets to the sun is 4.4 billion kilometers. Find the equation of Pluto's orbit assuming a center at (0,0). _____

2. An elliptically shaped garden is surrounded by a walkway. The garden is 20 meters long and 10 meters wide. The walk way is 2 meters wide. Find the area of garden and walkway. _____

3. Previous problem. What is the area of the walkway? _____

4. The Statuary Hall in the US Capitol is elliptical. It measures 14 meters wide and 30 meters long. If a person is standing at one focus, their whisper can be heard by a person standing at the other focus. How far apart are the two people? (Find the foci.) _____

Ch 6 Ls 1 How do hyperbolas find every point?   187

_____ #1 #2 ____/ 9 #3 ____/ 6 R ___/ 14 Total _____/ 29 _____
               Name   Checker

#1 1. What does a hyperbola look like? _____
   2. How does a hyperbola find each point? _____
   3. What equation do D and E use to find each point? _____
   4. Where is **A** and **B**? _____
   5. How is hyperbola different from ellipse? _____

#2 1. What equation do you use to find A in this hyperbola?

   ___ - ___ = 2___

   ___ = 2___

   2. You know A and D. Find E.   **A is 6. D is 2. What is E?**

   Solve it. What is E?
   ___ - ___ = 2___

   ___ - ___ = ___

   3. You know A and D. Find E.   **d is 7. e is 3. What is A?**

   Solve it. What is E?
   ___ - ___ = 2___

   ___ - ___ = ___

   4. You know A and D. Find E.   **d is 9. e is 4. What is A?**

   Solve it. What is E?
   ___ - ___ = 2___

   ___ - ___ = ___

#3 What equation do you use to find A in this hyperbola? Find the missing part.

Calculator? yes no

Find the missing part.

1.   d is 10, A is 1   d is 15, A is 3   d is 20, A is 4

    ___ - ___ = 2___   ___ - ___ = 2___   ___ - ___ = 2___

    ___ = ___   ___ = ___   ___ = ___

2.   d is 30, A is 4   d is 25, A is 3   d is 12, A is 3

    ___ - ___ = 2___   ___ - ___ = 2___   ___ - ___ = 2___

    ___ = ___   ___ = ___   ___ = ___

Review

1. What does a hyperbola look like? _____
2. Where does a hyperbola find each point? _____
3. What equation do D and E use to find each point? _____
4. Where is **A** in a hyperbola? _____
5. How is hyperbola different from ellipse? _____

6.   **d is 10, e is 4**   **d is 8, e is 2**   **d is 15, e is 3**

    ___ - ___ = 2___   ___ - ___ = 2___   ___ - ___ = 2___

    ___ = ___   ___ = ___   ___ = ___

7.   **d is 20, e is 4**   **d is 12, e is 3**   **d is 16, e is 6**

    ___ - ___ = 2___   ___ - ___ = 2___   ___ - ___ = 2___

    ___ = ___   ___ = ___   ___ = ___

8.   **d is 30, e is 5**   **d is 25, e is 7**   **d is 14, e is 4**

    ___ - ___ = 2___   ___ - ___ = 2___   ___ - ___ = 2___

    ___ = ___   ___ = ___   ___ = ___

Ch 6 Ls 2 Hyperbola Equation   189

_____ #1 #2 ____/ 7   #3 ____/ 8   R ____/ 4   Total ____/ 19 _____
           Name   Checker

#1 1. What is the circle formula? _____

2. How is ellipse formula different? _____

3. How is hyperbola different from ellipse equation? _____

4. What formula finds the points of a hyperbola? _____

#2 1. What's the 1st step to get a standard equation? $2x^2 - 4y^2 = 8$

Divide so _____. $\dfrac{2x^2}{} - \dfrac{4y^2}{} = 1$
What's the next step to a hyperbola?

Divide each fraction to get_____. $\dfrac{x^2}{} - \dfrac{y^2}{} = 1$
What's the last step?

$\dfrac{x^2}{} - \dfrac{y^2}{} = 1$

2. What's the 1st step to get a standard equation? $2x^2 - 8y^2 = 16$

Divide so _____. $\dfrac{2x^2}{} - \dfrac{8y^2}{} = 1$
What's the next step to a hyperbola?

Divide each fraction to get_____. $\dfrac{x^2}{} - \dfrac{y^2}{} = 1$
What's the last step?

$\dfrac{x^2}{} - \dfrac{y^2}{} = 1$

#3 What is happening? Describe each shape.

Calculator?
yes   no

1. $\dfrac{x^2}{5^2} - \dfrac{y^2}{3^2} = 1$ $\dfrac{x^2}{6^2} + \dfrac{y^2}{2^2} = 1$

_____   _____

_____   _____

2. $\dfrac{x^2}{9^2} - \dfrac{y^2}{2^2} = 1$ $x^2 + y^2 = 3^2$

_____   _____

_____   _____

3. $\dfrac{x^2}{8^2} - \dfrac{y^2}{3^2} = 1$ $\dfrac{x^2}{7^2} + \dfrac{y^2}{2^2} = 1$

_____   _____

_____   _____

4. $\dfrac{x^2}{6^2} - \dfrac{y^2}{1^2} = 1$ $\dfrac{x^2}{3^2} + \dfrac{y^2}{2^2} = 1$

_____   _____

_____   _____

Review   1. What is the circle formula? _____

2. How is ellipse formula different? _____

3. How is hyperbola different from ellipse equation? _____

4. What formula finds the points of a hyperbola? _____

Ch 6 Ls 3 How do hyperbolas make asymptotes?   191

_____ #1 #2 _____/10 #3 ____/ 4 R ____/11 Total ____/25 _____
     Name   Checker

#1 1. What formula puts A, B, and C together? _____
   2. What is the formula for asymptotes of a side to side hyperbola? _____
   3. How does asymptote equation change for up or down? _____
   4. What does a hyperbola use to show asymptotes? _____
   5. Draw A, B, and C. Where is the box? _____
   _____ Draw it.

#2 1. Write an equation.   a is 2, c is 3, what is b?

   What's the answer? _____

   _____

   2. Write an equation.   a is 2, c is 3, what is b?

   What's the answer? _____

   _____

   3. What are the 2 asymptotes? $\dfrac{x^2}{8^2} - \dfrac{y^2}{3^2} = 1$

   ___ ___

   4. What are the 2 asymptotes? $\dfrac{x^2}{7^2} - \dfrac{y^2}{1^2} = 1$

   ___ ___

   5. What are the 2 asymptotes? $\dfrac{x^2}{9^2} - \dfrac{y^2}{2^2} = 1$

   ___ ___

#3 You know 2 parts. Find the 3rd.s   Calculator?
                                                         yes   no

1. Write an equation.   a is 2, c is 3, what is b?   a is 5, c is 5, what is b?

   What's the answer?   _____   _____

                         _____   _____

2. Write an equation.   a is 3, c is 4, what is c?   a is 8, c is 10, what is b?

   What's the answer?   _____   _____

                         _____   _____

Review 1. What formula puts A, B, and C together? _____   Calculator?
        2. What is the formula for asymptotes of a side to side hyperbola? _____   yes   no
        3. How does asymptote equation change for up or down? _____
        4. What does a hyperbola use to show asymptotes? _____
        5. Draw A, B, and C. Where is the box? _____
        _____ Draw it.

What are the 2 asymptotes?

6. $\dfrac{x^2}{7^2} - \dfrac{y^2}{3^2} = 1$ $\dfrac{x^2}{8^2} - \dfrac{y^2}{2^2} = 1$ $\dfrac{x^2}{9^2} - \dfrac{y^2}{4^2} = 1$

7. $\dfrac{x^2}{5^2} - \dfrac{y^2}{2^2} = 1$ $\dfrac{x^2}{7^2} - \dfrac{y^2}{1^2} = 1$ $\dfrac{x^2}{8^2} - \dfrac{y^2}{3^2} = 1$

Ch 6 Ls 4 How do hyperbolas move?   193

_____ #1 #2 ____/ 9 #3 ____/ 10 R ____/ 4 Total ____/ 23 _____
Name   Checker

#1 1. How does circle formula change for moving from 0, 0? _____
   2. How does ellipse change for a different center? _____
   3. How do hyperbola's change for moving from 0, 0? _____
   4. What word helps to remember how the signs move? _____

#2 1. What is the eccentricity?   **A is 2.5   C is 8.5**

$$\text{eccentricity} = \frac{C \text{ is } \underline{}}{A \text{ is } \underline{}} \qquad \text{What is it}$$

$$\text{eccentricity} = \underline{}$$

2. How did the hyperbola move? $\dfrac{(x+3)^2}{9^2} - \dfrac{(y+4)^2}{7^2} = 1$

It moved _____

3. How did the hyperbola move? $\dfrac{(x-5)^2}{2^2} - \dfrac{(y-3)^2}{3^2} = 1$

It moved _____

4. How did the hyperbola move? $\dfrac{(x-2)^2}{3^2} - \dfrac{(y-3)^2}{4^2} = 1$

It moved _____

5. How did the hyperbola move? $\dfrac{(x+3)^2}{9^2} - \dfrac{(y+4)^2}{7^2} = 1$

It moved _____

#3 Find the eccentricity.

1. A is 1.5   C is 9.0   A is 1.2   C is 8.0   Calculator? yes   no

   Eccentricity is _____   Eccentricity is _____

2. A is 2.5   C is 7.5   A is 3.0   C is 12.0

   Eccentricity is _____   Eccentricity is _____

3. A is 1.7   C is 8.5   A is 1.2   C is 6.0

   Eccentricity is _____   Eccentricity is _____

4. A is 1.5   C is 11.5   A is 4.0   C is 14.0

   Eccentricity is _____   Eccentricity is _____

5. A is 2.7   C is 8.1   A is 3.4   C is 8.5

   Eccentricity is _____   Eccentricity is _____

Review   1. How does circle formula change for moving from 0, 0? _____   Calculator? yes   no

_____

2. How do hyperbola's change for moving from 0, 0? _____

3. What word helps to remember how the signs move? _____

How did these hyperbola move?

4. $\dfrac{(x-1)^2}{8^2} - \dfrac{(y+4)^2}{2^2} = 1$ $\dfrac{(x-2)^2}{9^2} - \dfrac{(y-3)^2}{1^2} = 1$

   It moves _____   It moves _____

5. $\dfrac{(x+3)^2}{7^2} - \dfrac{(y-2)^2}{3^2} = 1$ $\dfrac{(x-2)^2}{12^2} - \dfrac{(y+3)^2}{5^2} = 1$

   It moves _____   It moves _____

_____ #1 #2 #3 _____ / 16 Back _____ / 14 Total _____ / 30
Name

1. Hyperbola _____
2. Hyperbola Equation _____
3. Asymptotes Box _____
4. Hyperbola Not at 0 _____

#2 What equation do you use to find A in this hyperbola?

1. **d is 9, A is 1**   **d is 14, A is 2**   **d is 18, A is 4**

    ___ - ___ = 2 ___   ___ - ___ = 2 ___   ___ - ___ = 2 ___
    ___ = ___   ___ = ___   ___ = ___

2. **d is 30, A is 5**   **d is 28, A is 3.5**   **d is 13.8, A is 2.3**

    ___ - ___ = 2 ___   ___ - ___ = 2 ___   ___ - ___ = 2 ___
    ___ = ___   ___ = ___   ___ = ___

#3 Is each shape as circle, ellipse, or hyperbola?

1. $\dfrac{x^2}{5^2} - \dfrac{y^2}{3^2} = 1$ $\dfrac{x^2}{6^2} + \dfrac{y^2}{2^2} = 1$

   _____   _____

2. $\dfrac{x^2}{9^2} - \dfrac{y^2}{2^2} = 1$ $x^2 + y^2 = 3^2$

   _____   _____

3. $\dfrac{x^2}{8^2} - \dfrac{y^2}{3^2} = 1$ $\dfrac{x^2}{7^2} + \dfrac{y^2}{2^2} = 1$

   _____   _____

#4 You know 2 parts. Find the 3rd.   Calculator? yes   no

1. a is 7, c is 3, what is b?   a is 8, c is 5, what is b?

   Equation? _____   _____

   Answer? _____   _____

2. a is 9, c is 4, what is c?   a is 8, c is 10, what is b?

   Equation? _____   _____

   Answer? _____   _____

#5 What are the 2 asymptotes?   Calculator? yes   no

1. $\dfrac{x^2}{8^2} - \dfrac{y^2}{4^2} = 1$ $\dfrac{x^2}{6^2} - \dfrac{y^2}{2^2} = 1$ $\dfrac{x^2}{9^2} - \dfrac{y^2}{3^2} = 1$

   ▭ ▭   ▭ ▭   ▭ ▭

2. $\dfrac{x^2}{9^2} - \dfrac{y^2}{4^2} = 1$ $\dfrac{x^2}{7^2} - \dfrac{y^2}{2^2} = 1$ $\dfrac{x^2}{8^2} - \dfrac{y^2}{3^2} = 1$

   ▭ ▭   ▭ ▭   ▭ ▭

#6 How does each equation move?   Calculator? yes   no

1. $\dfrac{(x-1)^2}{8^2} - \dfrac{(y+4)^2}{2^2} = 1$ $\dfrac{(x-2)^2}{9^2} - \dfrac{(y-3)^2}{1^2} = 1$

   It moves _____   It moves _____

   _____   _____

2. $\dfrac{(x+3)^2}{7^2} - \dfrac{(y-2)^2}{3^2} = 1$ $\dfrac{(x-2)^2}{12^2} - \dfrac{(y+3)^2}{5^2} = 1$

   It moves _____   It moves _____

   _____   _____

Ch 7 Ls 1 How a parabola makes a focus.   197

_____ #1 #2 ____/ 6   #3 ____/ 4   R ____/ 4   Total ____/ 14 _____
         Name   Checker

#1 1. What formula finds the focus for a parabola? _____
2. Where does the directrix go? _____
3. What is the latus rectum? _____
4. What formula finds the latus rectum? _____

#2 1. Put A into the formula. $y = \frac{1}{2} x^2$
What equation is it?

Find the focus. Draw the parabola and directrix.

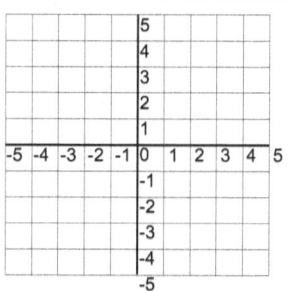

2. Put A into the formula. $y = \frac{1}{16} x^2$
What equation is it?

Find the focus. Draw the parabola and directrix. $\frac{1}{4}$

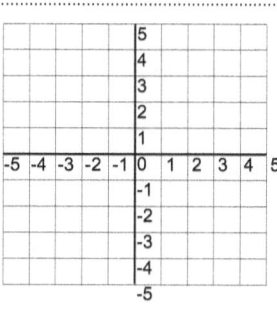

#3 Draw the parabola and directrix.   Calculator? yes no

1. $y = \frac{1}{2} x^2$ $y = \frac{1}{8} x^2$

   $\frac{1}{4}$ $\frac{1}{4}$

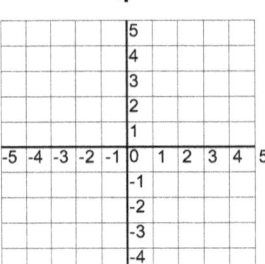
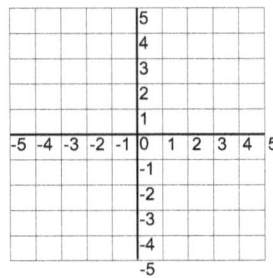

2. $y = 2x^2$ $y = 4x^2$

   $\frac{1}{4}$ $\frac{1}{4}$

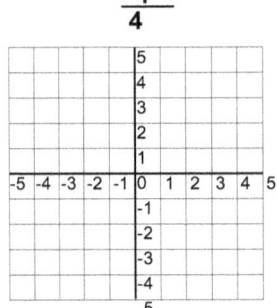
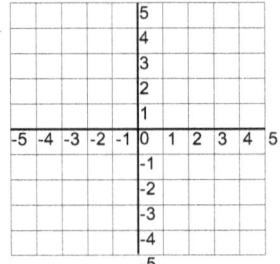

Review 1. What formula finds the focus for a parabola? _____

2. Where does the directrix go? _____

3. What is the latus rectum? _____

4. What formula finds the latus rectum? _____

Ch 7 Ls 2 How Quick Formula makes a parabola.   199

_____   #1 #2 ____/10   #3 ____/ 8   R _____/10   T _____/ 28 _____
            Name   Checker

#1 1. What is the quick formula? _____
   2. What part of it changes where the focus is? _____
   3. What does the A decide? _____
   4. Name 4 steps to change a trinomial to quick formula. _____
   _____

#2 1. How steep is it? How does it move? $y = (x - 3)^2 + 2$

   _____

   2. How steep is it? How does it move? $y = 2(x - 5)^2 + 4$

   _____

   3. How steep is it? How does it move? $y = 4(x - 4)^2 + 1$

   _____

   Make a quick equation for these.
   4. How steep is it? How does it move?   Slope: 4   Moves: -1, 2

   _____

   5. How steep is it? How does it move?   Slope: 5   Moves: 3, 4

   _____

   6. How steep is it? How does it move?   Slope: 4   Moves: -1, 2

   _____

#3 How stope is it? How does it move?   Calculator? yes no

1. $y = (x - 3)^2 + 2$ $y = 4(x - 2)^2 + 3$

vert ___, ___ faces _____   vert ___, ___ faces _____

2. $y = (x - 6)^2 + 4$ $y = 7(x - 2)^2 + 1$

vert ___, ___ faces _____   vert ___, ___ faces _____

3. $y = 9(x - 4)^2 + 5$ $y = 6(x - 3)^2 + 4$

vert ___, ___ faces _____   vert ___, ___ faces _____

4. $y = 7(x - 2)^2 + 3$ $y = 2(x - 4)^2 + 3$

vert ___, ___ faces _____   vert ___, ___ faces _____

Review 1. What is the quick formula? _____   Calculator? yes no

2. What part of it changes where the focus is? _____

3. What does the A decide? _____

4. Name 4 steps to change a trinomial to quick formula. _____
_____

Make a quick equation for these.

5. Slope: 2 Moves: -1, 2   Slope: 4 Moves: 3, -2

    _____   _____

6. Slope: 6 Moves: -5, 1   Slope: 5 Moves: 4, 2

    _____   _____

7. Slope: 8 Moves: 2, 2   Slope: 6 Moves: 1, -3

    _____   _____

Ch 7 Ls 3 How to change an equation to vertex equation.   201

_____ #1 #2 ____/10 #3 ____/ 8   R ____/10 T ____/ 28 _____
        Name   Checker

$$y = x^2 - 4x + 8 \qquad y = 2x^2 - 4x + 7$$

#1 1. What is the 1st thing with 1st equation? _____

2. What are 2 steps for complete the square? _____

3. How is the 2nd equation different? _____

4. Name 4 steps to change a trinomial to quick formula. _____

_____

#2 1. What stope is it? How does it move? $y = (x - 3)^2 + 2$

2. What stope is it? How does it move? $y = 2(x - 5)^2 + 4$

3. What stope is it? How does it move? $y = 4(x - 4)^2 + 1$

4. Make a quick equation for these.
   What stope is it? How does it move?   **Slope: 4   Moves: - 1, 2**

5. What stope is it? How does it move?   **Slope: 5   Moves: 3, 4**

6. What stope is it? How does it move?   **Slope: 2   Moves: - 3, 5**

#3 What stope is it? How does it move? Calculator? yes   no

1. $y = (x-3)^2 + 2$ $y = 4(x-2)^2 + 3$

2. $y = (x-6)^2 + 4$ $y = 7(x-2)^2 + 1$

3. $y = 9(x-4)^2 + 5$ $y = 6(x-3)^2 + 4$

4. $y = 7(x-2)^2 + 3$ $y = 2(x-4)^2 + 3$

Review
1. What is the quick formula? _____   Calculator? yes   no
2. What part of it changes where the focus is? _____
3. What does the A decide? _____
4. Name 4 steps to change a trinomial to quick formula. _____

Make a quick equation for these.

5. **Slope: 2   Moves: -1, 2**   **Slope: 4   Moves: 3, -2**

6. **Slope: 6   Moves: -5, 1**   **Slope: 5   Moves: 4, 2**

7. **Slope: 8   Moves: 2, 2**   **Slope: 6   Moves: 1, -3**

Ch 7 Ls 3 Substitute/Elimination Equations   203

_____ #1 #2 ____ / 8 #3 ____ / 3 R ____ / 7 Total ____ / 18 _____
Name   Checker

#1 1. What do you need to substitute equations? _____
2. How does the 1st equation substitute into the other? _____
3. After you find 1 variable, what happens? _____
4. Substitute it. What's the new equation? $x^2 - 1 = y$
   $-x + y = 1$

What's the 1st step?   _____

What's the answer?   _____

Use equation #2.   _____

Solve it. What's the point?   _____

_____

#2 1. How do you eliminate a variable? _____
2. What if no terms can be subtracted? _____
3. How do you subtract to eliminate a variable? _____
4. Eliminate it. What's the new equation? $x^2 - y = 9$
   $-x + y = 1$

What's the 1st step?   _____

What's the answer?   _____

Use equation #2.   _____

Solve it. What's the point?   _____

_____

#3
Substitute or eliminate it. Circle it.

1. A rocket is launched from the ground and follows a parabolic path represented by the equation $y = -x^2 + 12x$. At the same time, a flare is launched from a height of 4 meters and follows a straight path represented by the equation $y = -x + 12$. Graph the set. Do they intersect?

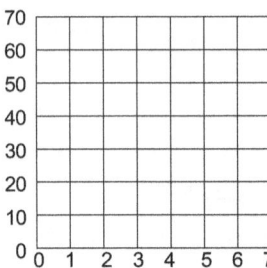

Calculator? yes   no

2. A pelican flying in the air over water drops a crab from a height of 10 m. The distance the crab is from the water as it falls can be represented by the function $y = -5t^2 + 10$, in sec. To catch the crab as it falls, a gull flies along a path represented by the function $y = -8t + 5$. Can the gull catch the crab before the crab hits the water?

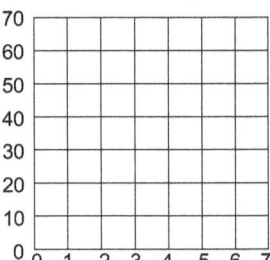

3. A beetle crawls at a certain speed and walks at another speed. If the beetle crawls for 10 seconds and walks for 9 sec it goes 85 m. If it crawls for 30 sec and walks for 2 sec, it travels 130 m. Find the beetle's crawling and walking speeds.

$10c + 9w = 85$
$30c + 2w = 130$

_____

_____

_____

_____

_____

Review
1. What do you need to substitute equations? _____
2. How does the 1st equation substitute into the other? _____
3. After you find 1 variable, what happens? _____
4. Name 3 reasons there is no answer. _____
_____
5. How do you eliminate a variable? _____
6. What if no terms can be subtracted? _____
7. How do you subtract to eliminate a variable? _____

Calculator? yes   no

Review Problems 205

_____ #1 #2 #3 _____ /14 #4 #5 _____ / 10 Total _____ / 24
               Name

1. Parabola Focus _____
2. Latus Rectum _____
3. Substitute Equation _____
4. Elimination Equation _____

#2 1.

Draw the parabola and directrix.

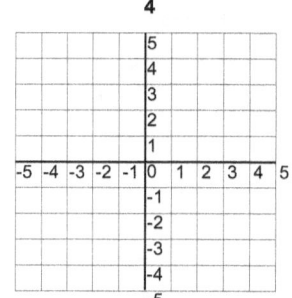
$y = \frac{1}{2} x^2$
$\frac{1}{4}$

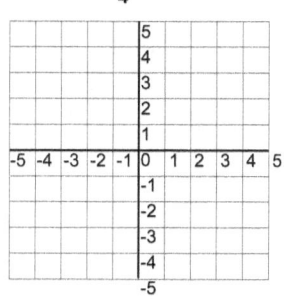
$y = \frac{1}{4} x^2$
$\frac{1}{4}$

#3 Where is the vertex? Which way does the parabola face?

1. $y = (x - 4)^2 + 2$ $y = 4(x - 5)^2 + 3$

    vert _____, _____ faces _____   vert _____, _____ faces _____

2. $y = (x - 7)^2 + 4$ $y = 7(x - 8)^2 + 2$

    vert _____, _____ faces _____   vert _____, _____ faces _____

3. $y = 9(x - 3)^2 + 2$ $y = 6(x - 4)^2 + 2$

    vert _____, _____ faces _____   vert _____, _____ faces _____

4. $y = 7(x - 6)^2 + 3$ $y = 2(x - 6)^2 + 3$

    vert _____, _____ faces _____   vert _____, _____ faces _____

#4 Make a quick equation for these.   Calculator? yes no

1. **Slope: 4   Moves: -1, 3**   **Slope: 0.5   Moves: 2, -2**
   _____   _____

2. **Slope: 7   Moves: -3, 2**   **Slope: 3   Moves: 6, 2**
   _____   _____

3. **Slope: 8   Moves: 3, 2**   **Slope: 6   Moves: 1, -4**
   _____   _____

#5 Substitute or eliminate it? What's the new equation?   Calculator? yes no

1. $4x^2 - 3y = 18$ $2x^2 + y = 6$
      $3x^2 + y = 7$ $2x^2 - y = 5$

What's the 1st step?   _____   _____

What's the answer?   _____   _____

Use equation #2.   _____   _____

Solve it.
What's the point?   _____   _____

2. $4x^2 + 6y = -9$ $4x^2 + 3y = 4$
      $2x^2 - 10y = -11$ $6x^2 - 6y = -1$

What's the 1st step?   _____   _____

What's the answer?   _____   _____

Use equation #2.   _____   _____

Solve it.
What's the point?   _____   _____

Ch 8 Ls 1 Fractions and Probability. 207

_____ #1 #2 ____/ 9 #3 ____/ 6 R ____/ 7 Total ____/ 22 _____
            Name   Checker

#1 1. How does a fraction measure an event? _____

2. How does the formula change to add total chances? _____

3. What is the difference between theoretical and experimental chances? _____
_____

4. How do you find the chances for 2 events happening? _____

5. Flip the coin 8 times. Get 3 tails. What's the equation? _____

#2 1. Make an equation.   A bag of candy has 5 green, 5 red, and 10 blue.
                                  What are the chances of picking the colors?

Change to percents.
                              _____

Jeff took all the blue ones.
What are the fractions now?   _____

Change to percents.
                              _____

                              _____

2. Here's 2 quizzes. What $\dfrac{27}{27 + 8}$
   are the percent scores?

                              ___ of ___ correct

3. Here's 2 quizzes. What $\dfrac{18}{18 + 2}$
   are the percent scores?

                              ___ of ___ correct

4. You have 4 coins and you want to find out the
   percent chance of getting 1 tail when you flip them.

                    _____   What percent
                    _____   is that?

It's _____ chances of getting 1 tail out of 4 coins.

#3 1.

What are the percent scores?

$$\frac{27}{27+6}$$ ___ of ___ correct ___% correct

$$\frac{42}{42+8}$$ ___ of ___ correct ___% correct

$$\frac{18}{18+2}$$ ___ of ___ correct ___% correct

Calculator? yes   no

2.

$$\frac{42}{42+8}$$ ___ of ___ correct ___% correct

$$\frac{67}{67+7}$$ ___ of ___ correct ___% correct

$$\frac{98}{98+2}$$ ___ of ___ correct ___% correct

Review

1. How does a fraction measure an event? _____

Calculator? yes   no

2. Flip the coin 8 times. Get 3 tails. What's the complement? _____

3. How does the formula change to add total chances? _____

4. What formula finds a chance of success for all chances? _____

5. How do you find all the possibilities when 3 coins are flipped? _____

6. Make an equation.

Change to percents.

A bag of candy has 8 green, 7 red, and 5 blue. What are the chances of picking each color?

Gr   **Red**   **Bl**

_____   _____   _____

JJ took all the blue ones. What are the fractions now?

Change to percents.

_____   _____   _____

_____   _____   _____

7. Make an equation.

Change to percents.

A bag of candy has 4 green, 6 red, and 10 blue. What are the chances of picking each color?

Gr   **Red**   **Bl**

_____   _____   _____

JJ took all the red ones. What are the fractions now?

Change to percents.

_____   _____   _____

_____   _____   _____

Ch 8 Ls 2 Independent/Dependent Events   209

_____ #1 #2 ____ / 9 #3 ____ /8 R ____ / 7 Total ____ / 24 _____
          Name   Checker

#1 1. What are independent events? _____

2. What are dependent events? _____

3. What does a factorial multiply? _____

4. How do you find all the possibilities when 3 coins are flipped? _____

_____

5. What is the Fundamental Counting Formula? _____

6. How do you find the average of 2 events happening? _____

#2 1. How do you solve this? **Mr B chooses 3 digits from 1 to 10 for his code, but he can't use a digit twice. How many possible codes?**

What if it's 1 digit and a letter?   _____

What if it's 2 digits and a letter?   _____

                                                    _____

2. How do you solve this? **Dr J's restaurant has 3 soups, 4 types of potato and 5 main meals. How many combinations are there?**

They ran out of a soup.
How many now?   _____

They have 2 desserts.
How many possibilities?   _____

                                                    _____

3. How do you solve this? **6 kids run for president, vice president, and secretary. How many possible ways can they be chosen?**

One more kid joins them.
How many possibilities?   _____

If they include treasurer,
how many possibilities?   _____

                                                    _____

#3

Dependent/Independent Events

1. Mr D chooses 3 digits from 1 to 20 for his code, but he can't use a digit twice. How many possible codes?

   Independent
   Dependent

2. Ten kids run for president, vice pres, and secretary. How many possible ways can they be chosen?

   Independent
   Dependent

3. A die is tossed twice. Find the probability of getting 1 or 2 on the 1st toss and 4, 5, or 6 on the 2nd toss.

   Independent
   Dependent

4. A bag contains 5 blacks, 3 white, and 2 red marbles. In each draw a marble is drawn and not replaced. In 3 draws, find the probability of getiing black, white, and red marbles in theat order.

   Independent
   Dependent

Dr J's restaurant has 2 soups, 3 types of potato and 5 main meals. How many combinations are there?

   Independent
   Dependent

Person A should be alive is 0.8 in 20 years and B is alive in 0.5 in 20 years. What's the chance that both are alive then?

   Independent
   Dependent

Two balls are drawn succesively without replacement from a box which contains 5 white and 4 black balls. Find the chance that the 1st ball is white and the 2nd is black.

   Independent
   Dependent

A deck of playing cards has two different colors and six different suits of 10 cards each. If four cards are dealt from the deck, what is the percent probability that all four are of the same color?

   Independent
   Dependent

Calculator?   yes   no

Review

1. What are independent events? _____

2. How do independent events find the number of possiblities? _____

3. What are dependent events? _____

4. How do dependent events find the number of possibilities? _____

5. What does a factorial multiply? _____

6. How do you find all the possibilities when 3 coins are flipped? _____

_____

7.   You have 5 coins and you want to find out the _____ percent chance of getting 1 tail when you flip them.____

   It's _____ chances of getting 1 tail out of 5 coins.

Ch 8 Ls 3 Permutation Formula   211

_____ #1 #2 ____/ 13 #3 ____/ 12 R ___/ 8 Total ____/ 33 _____
Name   Checker

#1 1. What does a permutation find? _____
2. How does a permutation count events? _____
3. How do you solve a permutation? _____
4. What is a permutation formula? _____
5. How do you find a permutation when parts are repeated? _____
_____

#2 1. What is happening?   P(5, 2)   P(5, 3)

Solve each permutation.   _____   _____

                          _____   _____

2. What is happening?   P(6, 2)   P(6, 3)

Solve each permutation.   _____   _____

                          _____   _____

3. What equation counts ways
these can be arranged?   **STOP**   **POPS**

Solve the equation.   _____   _____

                          _____   _____

4. What equation counts ways
these can be arranged?   **STOPS**   **ZOSOO**

Solve the equation.   _____   _____

                          _____   _____

#3 Solve these with permutions.   Calculator? yes   no

1. P(7, 2)   P(7, 3)   P(7, 4)

_____   _____   _____

2. P(8, 2)   P(8, 3)   P(8, 4)

_____   _____   _____

3. P(9, 2)   P(9, 3)   P(9, 4)

_____   _____   _____

4. P(10, 2)   P(10, 3)   P(10, 4)

_____   _____   _____

Review 1. How does a permutation count events? _____   Calculator? yes no

2. How do you solve a permutation? _____

3. What does a factorial do? _____

4. How do you find a permutation when parts are repeated? _____

_____

5. What equation counts ways these can be arranged?   **DROP**   **DROPPED**

_____   _____

Solve the equation.

_____   _____

6. What equation counts ways these can be arranged?   **LINEAR**   **EQUATE**

_____   _____

Solve the equation.

_____   _____

Ch 8 Ls 4   Combination Formulas   213

_____   #1 #2 ____/ 9   #3 ____/ 12   R ____/ 6   Total ____/ 27   _____
Name   Checker

#1 1. What makes a combination problem happen? _____

2. How does a group make a combination problem? _____

3. What is a combination formula? _____

4. How do you solve the combination C (4, 2)? _____

5. Why is **what's taken** important? _____

#2 1. What's the 1st step to solve each combination?   C(5, 2)

What's the 2nd step?   ___ x ___ = ___

___ ÷ ___ = ___

2. What's the 1st step to solve each combination?   C(6, 2)

What's the 2nd step?   ___ x ___ = ___

___ ÷ ___ = ___

3. What's the 1st step to solve each combination?   C(6, 3)

What's the 2nd step?   ___ x ___ = ___

___ ÷ ___ = ___

4. What equation solves this?   There's 5 students. 2 must sit upfront and the others in the back. How many ways can it happen?

Solve the equation.   _____

Make an equation if there are 6 students instead of 5.   _____

Solve the equation.   _____

_____

#3 Solve these with combinations.   Calculator?
                                                       yes   no

1. **C(7, 2)**   **C(7, 3)**   **C(7, 4)**

    _____   _____   _____

2. **C(8, 2)**   **C(8, 3)**   **C(8, 4)**

    _____   _____   _____

3. **C(9, 2)**   **C(9, 3)**   **C(9, 4)**

    _____   _____   _____

4. **C(10, 2)**   **C(10, 3)**   **C(10, 4)**

    _____   _____   _____

Review 1. What makes a combination problem happen? _____   Calculator?
                                                                            yes   no
2. How does a combination count events? _____

3. How do you solve the combination **C (4, 2)**? _____

4. Why is **what's taken** important? _____

5. **There's 5 students. 2 must sit upfront and the others in the back. How many ways can it happen?**   _____

_____

_____

_____

6. **Now there's 7 students and 3 must sit upfront. How many ways can it happen?**   _____

_____

_____

_____

Review Problems 215

_____ #1 #2 #3 _____ / 13 #4 #5 #6 _____ / 18 Total _____ / 31
Name

1. Fraction Probability _____
2. 2nd Formula _____
3. Factorial _____
4. Permutation Formula _____
5. Combinations Formula _____

#2 Fractions and Probability.

1.

$$\frac{17}{17+3} \qquad \frac{32}{32+8} \qquad \frac{23}{23+2}$$

What are the percent scores?

___ of ___ correct   ___ of ___ correct   ___ of ___ correct
____% correct   ____% correct   ____% correct

2.

$$\frac{46}{46+4} \qquad \frac{27}{27+6} \qquad \frac{18}{18+2}$$

___ of ___ correct   ___ of ___ correct   ___ of ___ correct
____% correct   ____% correct   ____% correct

#3 Independent Events

1. Mr D chooses 3 digits from 1 to 20 for his code, but he can't use a digit twice. How many possible codes? _____

What if it's 1 digit and a letter? _____
_____
_____

2. Dr J's restaurant has 2 soups, 3 types of potato and 5 main meals. How many combinations are there? _____

They ran out of a soup. How many now? _____

They have 2 desserts. How many possibilities? _____

#4 Solve with permutations.   Calculator? yes no

1. P(7, 2)   P(7, 3)   P(7, 4)

    _____   _____   _____

2. P(8, 2)   P(8, 3)   P(8, 4)

    _____   _____   _____

#5 Solve these with combinations.   Calculator? yes no

1. C(7, 2)   C(7, 3)   C(7, 4)

    _____   _____   _____

2. C(8, 2)   C(8, 3)   C(8, 4)

    _____   _____   _____

#6 Solve these story problems.   Calculator? yes no

1. How many possible words can be made?

   DROP   **DROPPED**

   _____ = ____   _____ = ____

2.
   LINEAR   **EQUATE**

   _____ = ____   _____ = ____

3. There's 10 students. 3 must sit upfront and the others in the back. How many ways can it happen?

   _____
   _____
   _____

4. TOoops!!! I miscounted. There's 11 students and 3 must upfront and the others in the back. How many ways can it happpen?

   _____
   _____
   _____

Ch 9 Ls 1 Mutually Inclusive/Exclusive Problems   217

_____ #1 #2 ____ / 8 #3 ____ / 10 R ___ / 4 Total ____ / 22 _____
        Name   Checker

#1 1. What does mutually exclusive mean? _____
        2. What does mutually inclusive mean? _____
        3. What is an overlap? _____
        4. Is the overlap added or subtracted? _____

#2 1. Find the odds of drawing a king   What are the   What are the
           and any heart from a deck of cards.   chances of a king?   chances of a heart?

           What finds the overlap?   ▭   ▭

           What are the chances?   _____

                                                   _____

        2. Find the odds of drawing a king   What are the   What are the
           and a jack from a deck of cards.   chances of a king?   chances of a jack?

           What finds the overlap?   ▭   ▭

           What are the chances?   _____

                                                   _____

                                    Is it exclusive or Inclusive?

        3. P(A) = 1/2 P(B) = 1/4 P(A or B) = ?   4. P(A) = 1/2 P(B) = 1/4 - P(A and B) 1/10 = ?

                                                                                          What's the
                Exclusive Inclusive?   Exclusive Inclusive?   answer?

            ▭ + ▭ = ▭   ▭ + ▭ - ▭ = ▭

#3
Solve these problems.

1. There are 4000 subscribers to an Internet service provider. Of these, 1200 own Micro computers, 2700 own a Peach computer, and 100 own both. What is the probability that a subscriber selected at random owns both?

    Exclusive Inclusive?   Calculator?
                                                 yes no

2. Zara has a stack of playing cards consisting of 12 hearts, 10 spades, and 8 clubs. If she selects a card at random from this stack, what is the probability that it is a heart or a club?

    Exclusive Inclusive?

3. You have a fair die. You throw it and get a 6. You throw it a 2nd time. What are the chances of getting a number 6 again.

    Exclusive Inclusive?

4. You have a fair die. You throw it and get a 5. You throw it a 2nd time. What are the chances of getting not getting a number 5.

    Exclusive Inclusive?

#4
Solve these problems.

1. $P(A) = 1/4$ $P(B) = 1/5$ $P(A \text{ or } B) = ?$

2. $P(A) = 1/3$ $P(B) = 3/5 - 1/10 = ?$

3. $P(A) = 1/2$ $P(B) = 1/4$ $P(A \text{ or } B) = ?$

4. $P(A) = 2/5$ $P(B) = 1/4 - 1/20 = ?$

5. $P(A) = 2/5$ $P(B) = 1/8$ $P(A \text{ or } B) = ?$

6. $P(A) = 3/7$ $P(B) = 1/2 - 3/14 = ?$

Review

1. What does mutually exclusive mean? _____

    Calculator? yes no

2. What does mutually inclusive mean? _____

3. What is an overlap? _____

4. Is the overlap added or subtracted? _____

Ch 9 Ls 2 Multiply Combinations Problems

_____ #1 #2 ____ / 7 #3 ____ / 2 R ____ / 5 Total ____ / 14 _____
　　　Name　　　　　　　　　　　　　　　　　　　　　　　　　　　　　　　　　　　Checker

#1 1. What kind of problem multiplies different combinations? _____

2. Why do fhey use combinations? _____

3. What are the 2 steps for solving them. _____

#2 1. What are the combinations?   A basket contains 4 apples, 5 bananas, and 8 plums. How many ways can 2 apples, 1 banana, and 2 plums be chosen?

$$C(\ ,\ ) \cdot C(\ ,\ ) \cdot C(\ ,\ )$$

_____ X _____ X _____ X _____　　Solve the combinations.

　　　　　　　　　　　　　　　　　Multiply the combinations. What's the answer?

There are _____ different ways to choose the fruit.

2. What are the combinations?   4 boys and 6 girls signed up for 4 person teams. Find the percent a boy will be on a team with 3 girls.

Solve the 1st 2 combinations.　　$$\frac{C(\ ,\ ) \cdot C(\ ,\ )}{C(\ ,\ )}$$

Multiply the 1st 2 combinations.　　_____　　_____

Find the bottom combination.　　　　　_____

Divide the 1st 2 by the total.　　　　　_____

Change to a percent.　　　　　　　　_____

#3 1. **5 red cards and 5 black cards are put in a bag. Find the chance that 2 red and 2 black cards will be selected from the bag.**   C(,) • C(,)   Calculator? yes no

   Solve the 1st 2 combinations. _____ _____

   _____ _____

   Multiply the 1st 2 combinations. _____

2. **This time 6 red cards and 6 black cards are put in a bag. Find the chance that 2 red and 2 black cards will be selected from the bag.**   C(,) • C(,)   Calculator? yes no

   Solve the 1st 2 combinations. _____ _____

   _____ _____

   Multiply the 1st 2 combinations. _____

Review 1. What kind of problem multiplies combinations? _____   Calculator? yes no

2. Where do the total parts go in each equation? _____

3. Where do the parts taken out of each combination go? _____

4. What does it divide by? _____

5. What are the 2 steps to find each set of combinations? _____

_____

6. **3 boys and 9 girls signed up for 4 person teams. Find the odds a boy will be on a team with 3 girls.** $\dfrac{C(,) \cdot C(,)}{C(,)}$

   Solve the 1st 2 combinations. _____ _____

   _____ _____

   Multiply the 1st 2 combinations. _____

   Find the bottom combination. _____

   Divide the 1st 2 by the total. _____

   Change to a percent. _____

Ch 9 Ls 3 Multiply Permutation Problems 221

_____ #1 #2 ____/ 9 #3 ____/ 6 R ___/ 5 Total ____/ 20 _____
Name   Checker

#1 1. What does Multiply Permutations do? _____
2. Why does it use permutations, not combinations? _____
3. Why are there 3 permutations multiplied instead of 2? _____
4. What is the 1st step to Multiply Permutations? _____
5. How does multiply permutations make an answer? _____

#2 1. What is the 1st step? **There's 5 rap CDs and 3 metal CDs you arrange by rank. How many different ways for each kind?**

What permutation shows changing shelves? _____

How do you solve each permutation? _____

Solve each permutation. _____

What happens if the shelves can be changed? _____

Find a final answer. _____

2. What is the 1st step? **There's 6 rap CDs and 4 metal CDs you arrange by rank. How many different ways for each kind?**

What permutation shows changing shelves? _____

How do you solve each permutation? _____

Solve each permutation. _____

What happens if the shelves can be changed? _____

Find a final answer. _____

#3 1. **There's 3 people out 7 and 2 out of 5 people working being selected for a special award. Order is important. How many ways can they be selected?**

        Solve each permutation. _____

_____

_____

Calculator?
yes   no

2. **A store owner picks from 3 of 7 meat products and 3 of 6 vegetable products to advertise. Order is important. How many ways can they be selected?**

        Solve each permutation. _____

_____

_____

3. **There's 6 rap videos and 3 metal videos to be arranged by rank. How many different ways for each kind?**

        Solve each permutation. _____

_____

_____

4. **There's 3 of 7 morning patients and 2 of 5 afternoon patients that need to reschedule. Rank is important. How many different ways can it happen?**

        Solve each permutation. _____

_____

_____

Review 1. What does Multiply Permutations do? _____

       2. Why does it use permutations, not combinations? _____

       3. Why are there 3 permutations multiplied instead of 2? _____

       4. What is the 1st step to Multiply Permutations? _____

       5. How does multiply permutations make an answer? _____

Calculator?
yes   no

Review Problems   213

_____ #1 #2 #3 ____ / 10 #3 #4 __ __ / 3 Total ____ / 13
Name

1. Exclusive Probability _____
2. Inclusive Probability _____
3. Multiply Combinations _____
4. Add Sets of Multiple Combinations _____
5. Multiply Permutations _____

#2 Use Inclusive Formula.

1. People from our school of 120 were surveyed. They found that 59 people recycle plastic and 41 recyle glass. 26 of those people recycle both. What percent of the people recycle both plastic and glass?

plastic ——— glass ———

_____
_____
_____

percent? _____

2. 200 students from our town were quizzed about futbol or basketball. 78 liked futbol best and 47 like basketball. 25 said they liked both. What percent said they liked futbol or basketball best?

futbol ——— basketball ———

_____
_____

percent? _____

#3 Use Multiply Combinations or Pernutations.

1. Zara wants to choose 2 of 8 apples and 3 of 7 watermelon to give away. Order not important. How many choices is that?

_____
_____
_____
_____

2. A team of 7 and 8 people need to choose a captain and co-captain. What are the chances for both of them to happen?

_____
_____
_____
_____

3. A survey worker has to visit 3 of 8 houses and 2 of 7 houses tomorrow morning. (order not important).

4. There are 6 conservative, 4 independent and 8 liberal speakers. 3 are chosen at random. Find the probability that 1 from each party is chosen.

5. There 7 countries from Asia and 5 countries from Africa that we want to visit. Unfortunately, we only have time to visit 3 from each trip. Which are the chances?

6. How many ways can 2 futbol players and 3 basketball players be selected from 6 futbol and 7 basketball players?

7. Chenwa is awesome at soccer. There are 3 teams out of 10 from the Indian Soccer League and 5 from 16 teams from the Chinese Soccer League he would want to play for. What are the chances?

Ch 10 Ls 1 Standard Deviation   225

_____ #1 #2 ____ / 9 #3 ____ / 6 R ____ / 9 Total ____ / 24 _____
            Name   Checker

#1 1. What does standard deviations do? _____
2. Name the 4 steps. _____ _____ _____ _____
3. What happens after the the 4 steps? _____
4. What step finds the limits? _____
5. What are the 1st 4 steps called? _____

#2 1. Find the 1st step.   5   5   11

1st Average is ____ ÷ ____ = ____   **2nd step?**

Subtract 1st average ____ ____ ____   **3rd step?**

Squared ____ ____ ____   **What's the 4th?**

2nd Average is ____ ÷ ____ = ____   **What happens next?**

Square root is ____   **What finds the limits?**

Add 7 + ____ = ____   Subtract 7 - ____ = ____

2. Find the 1st step.   7   7   16

1st Average is ____ ÷ ____ = ____   **2nd step?**

Subtract 1st average ____ ____ ____   **3rd step?**

Squared ____ ____ ____   **What's the 4th?**

2nd Average is ____ ÷ ____ = ____   **What happens next?**

Square root is ____   **What finds the limits?**

Add 7 + ____ = ____   Subtract 7 - ____ = ____

#3 Practice the 4 step sandwich.   Calculator? yes no

1.   **4 4 12**   **6 12 12**

1st Average: ____ Sub ____ ____ ____   1st Average: ____ Sub ____ ____ ____

Squared ___ ___ ___   2nd Av ____   Squared ___ ___ ___   2nd Av ____

2.   **5 6 11**   **10 11 8**

1st Average: ____ Sub ____ ____ ____   1st Average: ____ Sub ____ ____ ____

Squared ___ ___ ___   2nd Av ____   Squared ___ ___ ___   2nd Av ____

3.   **7 6 10**   **12 12 8**

1st Average: ____ Sub ____ ____ ____   1st Average: ____ Sub ____ ____ ____

Squared ___ ___ ___   2nd Av ____   Squared ___ ___ ___   2nd Av ____

Review 1. What does standard deviations do? _____   Calculator? yes no

2. Name the 1st 4 steps. _____ _____ _____ _____

3. What happens after the the 1st 4 steps? _____

4. What step finds the limits? _____

5. What are the 1st 4 steps called? _____

6.   **6 6 12**   **4 4 7**

1st Average: ____ Sub ____ ____ ____   1st Average: ____ Sub ____ ____ ____

Squared ___ ___ ___   2nd Av ____   Squared ___ ___ ___   2nd Av ____

St Dev: ____ Add: ____ Subtract: ____   St Dev: ____ Add: ____ Subtract: ____

Solve these standard deviations.   Is a score outside? Yes No   Is a score outside? Yes No

6.   **7 7 13**   **8 8 14**

1st Average: ____ Sub ____ ____ ____   1st Average: ____ Sub ____ ____ ____

Squared ___ ___ ___   2nd Av ____   Squared ___ ___ ___   2nd Av ____

St Dev: ____ Add: ____ Subtract: ____   St Dev: ____ Add: ____ Subtract: ____

Is a score outside? Yes No   Is a score outside? Yes No

Ch 10 Ls 2 Standard Deviation Story Problems

_____ #1 #2 ____/ 8   #3 ____/ 2   R ____/ 5   Total ____/ 15 _____
         Name   Checker

#1   1. What buttons get a calculator into statistics mode? _____
         2. What Greek letter adds the numbers? _____
         3. What buttons find the 2nd average? _____
         4. What button finds standard deviation? _____

#2 1. What's the 1st step?   Zara is selling her car, but she doesn't know what to ask.
                                  She finds 3 cars close to hers. Use standard deviations.

        *Use 21, not 000s.   Rs 21,000   Rs 16,000   Rs 20,000

        What's the next step?   1st Average is _____ ÷ ____ = ____

        What's the last of the sandwich?   Subtract 1st average ____ ____ ____

        What happens next?   Squared ____ ____ ____

        What steps find the limits?   2nd Average is _____ ÷ ____ = ____

                                         Square root is ____

                        Add ____ + ____ = ____   Subtract ____ - ____ = ____

        2. What's the 1st step?   Zara is also selling her house, but she doesn't know what to ask.
                                  She finds 3 houses close to hers. Use standard deviations.

        *Use 640 not 000s.   Rs 640,000   Rs 648,000   Rs 650,000

        What's the next step?   1st Average is _____ ÷ ____ = ____

        What's the last of the sandwich?   Subtract 1st average ____ ____ ____

        What happens next?   Squared ____ ____ ____

        What steps find the limits?   2nd Average is _____ ÷ ____ = ____

                                         Square root is ____

                        Add ____ + ____ = ____   Subtract ____ - ____ = ____

#3 Build an equation, then solve it.   Calculator?
                                                  yes   no

1. **Mr T manages a furniture store. He uses standard deviation of the 4 closest stores to set his own price on a chair.**

    Rs 2800   _____

    Rs 3200   _____

    Rs 2600   _____

    Rs 3000   _____

                   _____

                   _____

2. **The toothpaste department is tight. Myers Company looks for a place for their toothpaste. Use these 3 toothpastes to make a decision.**

    Rs 190   _____

    Rs 210   _____

    Rs 260   _____

                   _____

                   _____

Review 1. What kind of story problem uses standard deviation? _____   Calculator?
                                                                                          yes   no
2. What buttons get a calculator into statistics mode? _____

3. What Greek letter adds the numbers? _____

4. What buttons find the 2nd average? _____

5. What button finds standard deviation? _____

Ch 10 Ls 3 How Box and Whiskers use quartiles.   229

_____ #1 #2 ____ / 11 #3 ____ / 8 R ____ / 7 T ____ / 26 _____
                Name   Checker

#1 1. How do you divide 100 meters into quartiles? _____

2. How do you find 3 medians? _____

Find the   3. **2   3   4   8   9   11   24   7   8   10   12   14   15   24**
3 medians.

       ___ ___ ___   ___ ___ ___

4. What does Box and Whiskers do? _____

5. What's the 1st step to find what numbers are outside the set? _____

_____

6. What's the magic number and what happens to it? _____

7. What happens with the magic answer? _____

#2 1. What are the 3 medians?   **2   7   8   10   12   13   24**

What is the quartile range?   ___ ___ ___

What makes the quartile range?   ___ - ___ = ___

What decides the outliers?   ___ x 1.5 = ___

Decide what's outside the range.   ___ + ___ = ___   ___ - ___ = ___

_____

2. What are the 3 medians?   **4   8   9   10   12   13   20**

What is the quartile range?   ___ ___ ___

What makes the quartile range?   ___ - ___ = ___

What decides the outliers?   ___ x 1.5 = ___

Decide what's outside the range.   ___ + ___ = ___   ___ - ___ = ___

_____

#3 Use 4 steps to find Box and Whiskers.   Calculator? yes no

1. **1 7 8 10 12 14 26**   **4 10 11 14 15 18 30**

   3 medians ___ ___ ___   ___ ___ ___

   Find Qs. ___ - ___ = ___   ___ - ___ = ___

   Magic # ___ x 1.5 = ___   ___ x 1.5 = ___

   ___ + ___ = ___   ___ - ___ = ___   ___ + ___ = ___   ___ - ___ = ___

   _____   _____

2. **6 14 16 18 19 20 32**   **6 9 10 12 13 14 25**

   3 medians ___ ___ ___   ___ ___ ___

   Find Qs. ___ - ___ = ___   ___ - ___ = ___

   Magic # ___ x 1.5 = ___   ___ x 1.5 = ___

   ___ + ___ = ___   ___ - ___ = ___   ___ + ___ = ___   ___ - ___ = ___

   _____   _____

Review 1. How do you divide 100 meters into quartiles? _____   Calculator? yes no

2. How do you find 3 medians? _____

3. **5 7 8 10 11 14 25**   **9 10 13 14 17 18 29**

   ___ ___ ___   ___ ___ ___

4. What does Box and Whiskers do? _____

5. What's the 1st step to find what numbers are outside the set? _____
   _____

6. What's the magic number and what happens to it? _____

7. What happens with the magic answer? _____

Ch 10 Ls 4 Box and Whiskers Practice   231

_____ #1 #2 ____ / 9 #3 ____ / 12 R ____ / 3 Total ____ / 24 _____
          Name   Checker

#1   1. Name 4 steps to Box and Whiskers. _____
_____
    2. How are Box/Whiskers the same as Standard Deviations? _____
_____
    3. Why is Standard Deviations more accurate? _____

#2   1. What are the 3 medians?   5   10   11   12   14   15   28

    What is the quartile range?   ___   ___   ___

    What makes the quartile range?   ___ - ___ = ___

    What decides the outliers?
                                         ___ x 1.5 = ___

    Circle what's outside the range.
                                         ___ + ___ = ___   ___ - ___ = ___

    2. What are the 3 medians?   12   18   21   23   24   28   35

    What is the quartile range?   ___   ___   ___

    What makes the quartile range?   ___ - ___ = ___

    What decides the outliers?
                                         ___ x 1.5 = ___

    Circle what's outside the range.
                                         ___ + ___ = ___   ___ - ___ = ___

    3. What are the 3 medians?   20   27   33   35   37   41   60

    What is the quartile range?   ___   ___   ___

    What makes the quartile range?   ___ - ___ = ___

    What decides the outliers?
                                         ___ x 1.5 = ___

    Circle what's outside the range.
                                         ___ + ___ = ___   ___ - ___ = ___

#3 The planets of our solar system take this long to make a length day. (compared to earth's hours).

Calculator? yes no

Story Problems.
5,832 h   1,406 h   153 h   20 h   17 h   11 h   10 h
venus   mercury   pluto   uranus   neptune   saturn   jupiter

1. Which planets are the quartles? _____
2. Subtract the 1st and 3rd quartile. _____
3. Multiply the magc number. _____
4. Which planets would create outliers? _____

Ojas has 7 grades for the science final. Are any of them outliers? (The grades are in order for the problem.)

70   80   85   85   85   90   90

5. Which grades are the quartles? _____
6. Subtract the 1st and 3rd quartile. _____
7. Multiply the magc number. _____
8. Which grades would create outliers? _____

Mr T is the science teacher and he keeps a log of the average science grades. Here's the average. (The grades are in order for the problem.)

65   70   75   80   80   85   85

9. Which grades are the quartles? _____
10. Subtract the 1st and 3rd quartile. _____
11. Multiply the magc number. _____
12. Which grades would create outliers? _____

Review   1. Name 4 steps to Box and Whiskers. _____
_____

2. How are Box/Whiskers the same as Standard Deviations? _____
_____

3. Why is Standard Deviations more accurate? _____

_____ #1 #2 _____ / 20 #3 #4 _____ / 12 Total _____ / 32
                Name

1. Standard Deviations _____
2. ASSA _____
3. Box and Whiskers _____
4. Quartile _____

#2   1.   **15   15   26**   **20   20   8**

1st Average: ____ Sub ____ ____ ____   1st Average: ____ Sub ____ ____ ____

Squared ____ ____ ____   2nd Av ____   Squared ____ ____ ____   2nd Av ____

St Dev: ____ Add: ____ Subtract: ____   St Dev: ____ Add: ____ Subtract: ____

Solve with   Is a score outside? Yes No   Is a score outside? Yes No
standard
deviation

     2.   **10   10   22**   **30   30   12**

1st Average: ____ Sub ____ ____ ____   1st Average: ____ Sub ____ ____ ____

Squared ____ ____ ____   2nd Av ____   Squared ____ ____ ____   2nd Av ____

St Dev: ____ Add: ____ Subtract: ____   St Dev: ____ Add: ____ Subtract: ____

          Is a score outside? Yes No   Is a score outside? Yes No

     3.   **40   40   11**   **35   35   20**

1st Average: ____ Sub ____ ____ ____   1st Average: ____ Sub ____ ____ ____

Squared ____ ____ ____   2nd Av ____   Squared ____ ____ ____   2nd Av ____

St Dev: ____ Add: ____ Subtract: ____   St Dev: ____ Add: ____ Subtract: ____

          Is a score outside? Yes No   Is a score outside? Yes No

     4.   **50   50   11**   **45   45   27**

1st Average: ____ Sub ____ ____ ____   1st Average: ____ Sub ____ ____ ____

Squared ____ ____ ____   2nd Av ____   Squared ____ ____ ____   2nd Av ____

St Dev: ____ Add: ____ Subtract: ____   St Dev: ____ Add: ____ Subtract: ____

          Is a score outside? Yes No   Is a score outside? Yes No

#3 1. **12 20 24 28 34 11 24**   **45 60 64 65 70 72 90**   Calculator?

Use Box and   3 medians   ____ ____ ____   ____ ____ ____   yes   no
Whiskers.
             Find Qs.   ____ - ____ = ____   ____ - ____ = ____

             Magic #   ____ x 1.5 = ____   ____ x 1.5 = ____

      ____ + ____ = ____   ____ - ____ = ____   ____ + ____ = ____   ____ - ____ = ____

      _____   _____

2. **30 45 46 48 50 51 68**   **44 65 67 70 72 74 98**

             3 medians   ____ ____ ____   ____ ____ ____

             Find Qs.   ____ - ____ = ____   ____ - ____ = ____

             Magic #   ____ x 1.5 = ____   ____ x 1.5 = ____

      ____ + ____ = ____   ____ - ____ = ____   ____ + ____ = ____   ____ - ____ = ____

      _____   _____

#4 1. 7 students made the following scores on a science exam:

   56 80 84 87 91 92 97

   Are any of them outliers for Box and Whiskers?

             ____ ____ ____

             ____ - ____ = ____

             ____ x 1.5 = ____

      ____ + ____ = ____   ____ - ____ = ____

      _____

2. 7 students made the following scores on a math exam:

   64 70 76 80 84 84 90

   Are any of them outliers for Box and Whiskers?

             ____ ____ ____

             ____ - ____ = ____

             ____ x 1.5 = ____

      ____ + ____ = ____   ____ - ____ = ____

      _____

Ch 11 Ls 1 Margin of Error 235

_____ #1 #2 ____/10 #3 ____/ 5 R ___/ 6 Total ____/ 21 _____
   Name   Checker

#1 1. What is a sample group? _____

2. Name 3 things that can decide a sample group. _____

_____

3. Is this Simple Random   There are 200 employees of a company. They put all
   Sampling?   their names in a hat and chose out 20 for a meeting.

          YES   NO   Why is it or isn't it? _____

_____

#2 1. What is the first step to the formula for Margin of Error? _____

2. What is the rest of the Margin of Error formula? _____

3. What happens to the percent results? _____

4. What does 95% certainty tell of? _____

5. What is the formula for Margin of Error? _____

6. Do this problem.   We asked 300 people if proathletes were overpaid.
   2 points   80% said they were. What is the Margin of Error fomula?

What is n and p? $ME = 2\sqrt{\dfrac{(1 - p)}{n}}$ Put the numbers in the equation.

$ME = 2\sqrt{\dfrac{(1 - 0.80)}{300}}$ Subtract P.

$ME = 2\sqrt{\dfrac{(0.20)}{300}}$ Multiply the numerator.

$ME = 2\sqrt{\dfrac{}{300}} = 2\sqrt{}$ Find the square root.

2 x _____ = _____ or _____ %

80% + 4%
80% − 4%   Between _____ % and _____ % believe proathletes are over paid.

#3 **1.** We asked 400 people if universities are too expensive. 90% said they were. Here's part of the Margin of Error formula.   Calculator? yes no

Do this Margin of error problem.
2 points

What is n and p? $ME = 2\sqrt{\dfrac{(1-p)}{}}$

Put the numbers in the equation. $ME = 2\sqrt{\dfrac{(1-)}{}}$

Subtract P. $ME = 2\sqrt{\dfrac{()}{}}$

Multiply the numerator. $ME = 2\sqrt{\dfrac{}{}} = 2\sqrt{}$

Find the square root. $2 \times \underline{} = \underline{}$ or $\underline{}\%$

90% + ___%
90% − ___%   Between ____% and ____% believe proathletes are over paid.

2. Is this Simple Random Sampling? Why?   Obtaining a list of boys first names by writing down names from your cricket team.

1 point   **YES   NO**   _____

3. Is this Simple Random Sampling? Why?   Finding the heights of the girls in your music class to decide the average height of girls in your school.

YES   NO   _____

4. Is this Simple Random Sampling? Why?   Taking a poll from your bus ride to school to find out how many students would go to a car wash event.

YES   NO   _____

Review 1. What is a sample group? _____   Calculator? yes no
2. Name 3 things that can decide a sample group. _____
3. What is the first step to the formula for Margin of Error? _____
4. What is the rest of the Margin of Error formula? _____
5. What happens to the percent results? _____
6. What is the formula for Margin of Error? _____

Ch 11 Ls 2 Central Tendency/Measures of Dispersion 237

_____ #1 #2 ____/10 #3 ____/ 4 R ___/ 7 Total ____/ 21 _____
           Name   Checker

#1 1. What finds the mean of a group of numbers? _____
    2. What does this letter stand for? Σ _____
    3. What finds the median of a group of numbers? _____
    4. What does mode find? _____
    5. **14, 17, 12, 19, 12, 20, 23**   What is the mean of these numbers?

                    Add them _____ Divide by _____ is _____
            6. What is the median of these numbers?

                        _____ is the median
            7. What is the mode of these numbers?

                        _____ is the mode

#2 1. What does Measure of Dispersion find? _____
    2. How do you find Range? _____
    3. Name 4 ways to find the Measure of Dispersion. _____
       _____
    4. Name the standard deviation code and what changes it. _____
       _____
    5. A company of 4 people ages are **24, 32, 33, and 35**. Find the standeard deviation.
       The 1st average is 31. What is the 2nd step?

    Subtract ____ ____ ____ ____   What's the 3rd step?
    Squared ____ ____ ____ ____   What's the 4th step?
    Average is _____   **Square Root?**
    Square Root is _____   Is 24 inside or outside?
                Circle one.   **Inside   outside**

#3   1. Candy bars are popular with the public. What are the mean, median,   Calculator?
     and mode? **Rs 30, 35, 30, 28, 30, 29, 40**   yes   no

           Mean _____   Median _____   Mode _____

   2. Jeans are popular with those not in the country. What are the mean, median,
      and mode? **Rs 370, 390, 399, 410, 410, 440, 600**

           Mean _____   Median _____   Mode _____

   3. Cars are popular with those who can afford them. What are the mean, median,
      and mode? **Rs 50, 70, 75, 80, 81, 90, 150**

           Mean _____   Median _____   Mode _____   Cars are in thousands.

   4. A company of 4 people ages are **24, 32, 33, and 35**. Find the standeard deviation.
      The 1st average is **31**. What is the 2nd step?

   Subtract ____ ____ ____ ____   What's the 3rd step?

   Squared ____ ____ ____ ____   What's the 4th step?

   Average is _____   **Square Root?**

   Square Root is _____   **Is 24 inside or outside?**

                Circle one.   **Inside   outside**

Review 1. What is a sample group? _____   Calculator?
        2. Name 3 things that can decide a sample group. _____   yes   no
        3. What is the first step to the formula for Margin of Error? _____
        4. What is the rest of the Margin of Error formula? _____
        5. What happens to the percent results? _____
        6. What is the formula for Margin of Error? _____
        7. Name the standard deviation code and what changes it. _____
        _____

Ch 11 Ls 3 Normal Distribution 239

_____ #1 #2 ____/ 10 #3 ____/ 6 R ____/ 5 Total ____/ 21 _____
           Name   Checker

#1 1. What kind of a graph does normal distribution use? _____

2. What is the name for this style of graph? _____

3. What is the middle called? _____

4. How much of the graph is within 1 standard deviation? _____

5. How much of the graph is in 2 standard deviations? _____

#2 1. On a standardized test Zara got a score of 78, which was exactly 1 standard deviation above the mean. If the standard deviation was 6, what was the mean score for the test?   A. 70   B. 84   C. 73   D. 72

_____

2. On a standardized test, the distribution of scores is normal, the mean of the scores is 76, and the standard deviation is 5.9. If a student scored 83, the student's score ranks
(1) below the 75th percentile   (3) between the 75th and 84th percentiles
(2) above the 95th percentile   (4) between the 84th and 95th percentiles

_____

3. A set of data with a mean of 55 and a standard deviation of 7.2 is normally distributed. Find +1 and + 2 standard deviation from the mean.

_____   _____

4. The mean score on an exam is 110 with a standard deviation of 13. Which score would be expected to occur less than 5% of the time?
(1) 95 (2) 108 (3) 140 (4) 130

_____

5. On a standardized test, Amav scored 112 exactly two standard deviations above the mean. If the standard deviation for the test is 8, what is the mean score for the test?

_____

#3   Normal Distribution   Calculator?
                                                                  yes   no

1. **Entry to a national university is determined by a national test. The scores on this test are normally distributed with a mean of 600 and a standard deviation of 100. Ojas wants to be admitted to this university and he knows that he must score better than at least 75% of the students who took the test. Ojas takes the test and scores 750. Will he be admitted to this university?**

   Circle one. **Yes   No**   Ojas scored in what
                                     standard deviation? _____

2. **A set of scores with a normal distribution has a mean of 60 and a standard deviation of 7. Approximately what percent of the scores fall in the range 46-74?**

   _____

3. **The time taken to assemble a car in a plant is a normal distribution of 24 hours and a standard deviation of 2.5 hours. What is the probability that a car can be assembled at this plant in a period of time less than 21 hours?**

   _____

4. **On a standardized test, the mean was 65 and the standard deviation was 6. What is the best approximation of the percent of scores that fell between 59 and 71?**

   _____

5. **The mean of a normally distributed set of data is 5 and the standard deviation is 5. Approximately 95% of all the cases will lie between which measures?**

   _____

6. **On a standardized test, the mean was 70 and the standard deviation was 7. What is the best approximation of the percent of scores that fell between 56 and 84?**

   _____

Review 1. What kind of a graph does normal distribution use? _____   Calculator?
        2. What is the name for this style of graph? _____   yes   no
        3. What is the middle called? _____
        4. How much of the graph is within 1 standard deviation? _____
        5. How much of the graph is in 2 standard deviations? _____

Ch 11 Ls 4 Binomial Theorem   241

_____ #1 #2 ____/ 8 #3 ____/ 3 R ____/ 4 Total ____/ 15 _____
        Name   Checker

#1 1. What does 1 more term mean? $(a + b)^2$ _____

2. What do the exponents add to? _____

3. How do coefficients get the next level? _____

4. What are the 3 rules that show how binomials multiply out? _____

_____

#2 1. Write $(a - b)^2$ in expanded form. $(a - b)^2$

Finish it. Estimate it to a 10th.   _____

Describe each graph.   _____

2. Write $(a - b)^3$ in expanded form. $(a - b)^3$

Finish it. Estimate it to a 10th.   _____

Describe each graph.   _____

3. Write $(a - b)^4$ in expanded form. $(a - b)^4$

Finish it. Estimate it to a 10th.   _____

Describe each graph.   _____

4. Write $(a - b)^5$ in expanded form. $(a - b)^5$

Finish it. Estimate it to a 10th.   _____

Describe each graph.   _____

#3 Use Bnomial Distributon to answer these.   Calculator?
                                                      yes   no

1. A coin is tossed 6 times. Calculate   B(6, 0.5)   p = 0.5   1 - p = 0.5
   the probability of obtaining more
   heads than tails.

   _____

   _____

   _____

2. JJ makes 2 fifths of his free throws.   _____
   He made 3 out of 4 in the next game.
   What's the probability for that?   _____

   Made $\frac{2}{5}$ Missed $\frac{3}{5}$ _____

   Made $\frac{3}{4}$ Missed $\frac{1}{4}$ _____

3. JJ makes 4 fifths of his free throws.   _____
   He made 3 out of 6 in the next game.
   What's the probability for that?   _____

   Made $\frac{4}{5}$ Missed $\frac{1}{5}$ _____

   Made $\frac{2}{6}$ Missed $\frac{4}{6}$ _____

Review 1. What does 1 more term mean? $(a + b)^2$ _____   Calculator?
        2. What do the exponents add to? _____   yes   no
        3. How do coefficients get the next level? _____
        4. What are the 3 rules that show how binomials multiply out? _____

        _____

Ch 11 Ls 5 Scatter Plots 243

_____ #1 #2 ____ / 7 #3 ____ /14 R ____ / 2 Total ____ / 23 _____
         Name   Checker

#1 1. What are the 4 main graphs? _____
_____

2. What are the positive graphs?

The 1st graph is positive and
the 2nd one is weak positive.

3. What are the negative graphs?

The 1st graph is negative and
the 2nd one is weak negative.

4. What is the other graph?

The _____
         graph.

#2 1. Decide the   Your free time decreases as the number of sports
        correlation.   you play increases; the data has a ? correlation.

_____

2. As a child gets older his weight increases has ? correlation?

_____

3. The price is not affected by the color of its zipper so the data has ? correlation.

_____

#3 Do the data sets have a positive, a negative, or no correlation?   Calculator?
correlation.   yes   no

1. Height versus weight of 50 male elephants.   _____

2. Distance driven versus gas used.   _____

3. Amount of time spent studying versus grades.   _____

4. Hair length versus height of 60 adult women.   _____

5. Distance walked in a pair of shoes vs thickness of the sole.   _____

6. The hours a jet is in flight and the number of miles flown.   _____

7. The hours in flight and the number of passengers.   _____

8. The hours in flight and the gallons of fuel remaining.   _____

9. The size of jar of baby food, the number jars a baby will eat.   _____

10. The speed of a runner and the number of races he wins.   _____

11. The size of a person and the number of fingers he has.   _____

12. The size of a bird and how much he will eat.   _____

13. The height of a roller coaster and how fast it falls.   _____

14. Age of a cell phone and number of features it has.   _____

Review 1. What are the 4 main graphs? _____   Calculator?
_____   yes   no

2. What is the other graph? _____

Ch 11 Ls 6 Scatter Plots/Linear Programming 245

_____ #1 #2 ____ / 6 #3 ____ / 1 R ____ / 4 Total ____ / 11 _____
           Name   Checker

#1 1. How many equations does the 1st Linear Programing use? _____

2. How does it find an answer? _____

3. How many equations does the 2nd Linear Programing use? _____

4. How does it find an answer? _____

#2 1. What is the grahing equations? $x \geq 1$ $y \geq 1$ $x + y \leq 6$

Graph the equations. $y \leq -x + 6$

2. What is the grahing equations? $1.5x + y \leq 5$ $x + y \leq 7$

Graph the equations. $y \leq -1.5x + 5$ $y \leq -x + 7$

#3. 1. **The bike company makes two styles of bicycles: #1, which sells for Rs 600, and #2, which sells Rs 1200. Each bicycle has the same frame and tires, but the assembly and painting time required for #1 is only 1 hour, while it is 3 hours for #2. There are 300 frames and 360 hours of labor available for production. How many bicycles of each model should be produced to maximize revenue?**

Calculator? yes   no

How many frames are there?

There are _____ of them. $x + y \leq$ _____

What is the equation for graphing?

$y \leq -x +$ _____

How many hours does it take for each?

$x + 3y \leq$ _____

$3y \leq -x +$ _____

$y \leq -x/3 +$ _____/3

What is the revenue?

$y < -x/3 +$ _____

$600x + 1200y$

Graph the 2 equations.

They should make _____ #1 bikes

and _____ #2 bikes.

Review 1. How many equations does the 1st Linear Programing use? _____   Calculator? yes   no

2. How does it find an answer? _____

3. How many equations does the 2nd Linear Programing use? _____

4. How does it find an answer? _____

Review Problems 247

_____ #1 #2 ____/ 13 #3 #4 ____/ 10 Total ____/ 23
            Name

1. Normal Distribution _____

2. Random Sampling Error _____

3. Normal Distribution _____

4. Binomial Theorem _____

4. Scatter Plots _____

5. Linear Programing _____

#2 1. The mean score on a normally distributed exam
       is 46 with a standard deviation of 12. Which score   _____
       would be expected to occur 2x standard deviation?

    2. 185 students took a math exam. The scores have
       a mean of 72 and a standard deviation of 9. How   _____
       many students in the class can be expected to
       receive a score between 82 and 90?

    3. Ojas put tags on 20 eagles and released them. Later,
       he catches 90 eagles; 16 eagles were tagged. What   _____
       is the best estimate for the Eagle population?

    4. Students took a test in math and the final grades   1x _____
       have a mean of 72 and a standard deviation of 10.
       What score received 1x and 2x standard deviation?   2x _____

    Are the following simple random samples or not? Why?

    5. Obtaining a list of boys first names by writing   Yes   No _____
       down names from your cricket team.
                                                               _____

    6. Finding the heights of the girls in your art   Yes   No _____
       class to decide the average height of girls in
       your school.   _____

    7. Take a poll from your bus ride to school   Yes   No _____
       to find out how many students would go to
       a car wash event.   _____

    8. Find an average height by taking all of the   Yes   No _____
       names and randomly taking 10% of them
       for the average.   _____

#3 1. Bubble Gum is popular with the public. What are the mean, median, and mode? **Rs 25, 28, 30, 30, 35, 37, 40**   Calculator? yes no

   Mean _____   Median _____   Mode _____

2. These shirts are popular with those not in the country. What are the mean, median, and mode? **Rs 370, 390, 399, 410, 410, 440, 600**

   Mean _____   Median _____   Mode _____

3. Trucks are popular with those who can afford them. What are the mean, median, and mode? **Rs 60, 68, 80, 90, 99, 120, 170**   Trucks are in thousands.

   Mean _____   Median _____   Mode _____

4. On a standardized test, the distribution of scores is normal, the mean of the scores is 135, and the standard deviation is 9.5. If a student scored 142, the student's score ranks
   (1) below the 75th percentile   (3) between the 75th and 84th percentiles
   (2) above the 95th percentile   (4) between the 84th and 95th percentiles   _____

5. A set of data with a mean of 64 and a standard deviation of 8.4 is normally distributed. Find +1 and + 2 standard deviation from the mean.   _____

#4   Is it positive (P), negative (N), or no correlation (C)?

1. Your free time decreases as the number of jobs you work increases; the data has a _____ correlation.

2. As a child gets older his weight increases has _____ correlation?

3. The price is not affected by the color of its zipper so the data has _____ correlation.

4. A student's class size determines their grade for the class _____ correlation?

5. A student's size does not effect their play for soccer _____ correlation.

Graph these linear graphs.

$1.5x + y \leq 5$

$x + y \leq 7$

$x \geq 1$

$y \geq 1$

Ch 12 Ls 1 Arithmetic Sequence Formula 249

_____ #1 #2 ____/ 11 #3 ____/ 9 R ____/ 17 T ____/ 37 _____
         Name   Checker

#1 1. What does a sequence find? _____
  2. What is the 1st number called? _____
  3. What is the distance between numbers called? _____
  4. What is the arithmetic sequence formula? _____
  5. How do you remember it? _____

#2 1. What is the 1st term and step?   **1, 4, 7, 10**

  1st term is _____   Step is _____   Find the next 2 terms.

  ____ ____

  2. What is the 1st term and step?   **2, 6, 10, 14**

  1st term is _____   Step is _____   Find the next 2 terms.

  ____ ____

  3. What is the 1st term and step?   **10, 16, 22, 28**

  Find the next 2 terms.   1st term is _____   Step is _____

  ____ ____

  4. Mentally find these.   1st term 5, step 4. What's the 4th term?

  ____ ____ ____ ____

  5. Mentally find these.   1st term 3, step 5. What's the 5th term?

  ____ ____ ____ ____ ____

  6. Mentally find these.   1st term 8, step 2. What's the 6th term?

  ____ ____ ____ ____ ____ ____

#3 Decide the step. Find the missing term.   Calculator? yes no

1.   **1, 4, ?, 10**   **4, 9, ?, 19**   **2, ?, 7, 10**

     Step ___ 1, 4, ___, 10   Step ___ 4, 9, ___, 19   Step ___ 1, ___, 7, 10

2.   **1, ?, 9, ?**   **6, ?, 14, ?**   **11, ?, 17, ?**

     Step ___ 1, ___, 9, ___   Step ___ 1, ___, 14, ___   Step ___ 1, ___, 7, ___

3.   **23, ?, 29, ?**   **35, ?, 41, ?**   **53, ?, 67, ?**

     Step ___ 23, ___, 29, ___   Step ___ 35, ___, 41, ___   Step ___ 53, ___, 67, ___

Review   1. What does a sequence find? _____   Calculator? yes no

      2. What is the 1st number called? _____

      3. What is the distance between numbers called? _____

      4. What is the arithmetic sequence formula? _____

      5. How do you remember it? _____

6.   1st term 1, step 2.   1st term 5, step 4.   1st term 8, step 5.
    What's the 4th term?   What's the 5th term?   What's the 6th term?

    ___ ___ ___ ___   ___ ___ ___ ___ ___   ___ ___ ___ ___ ___ ___

7.   1st term 10, step 3.   1st term 16, step 7.   1st term 11, step 4.
    What's the 4th term?   What's the 5th term?   What's the 6th term?

    ___ ___ ___ ___   ___ ___ ___ ___ ___   ___ ___ ___ ___ ___ ___

8.   1st term 14, step 8.   1st term 16, step 10.   1st term 21, step 5.
    What's the 4th term?   What's the 5th term?   What's the 6th term?

    ___ ___ ___ ___   ___ ___ ___ ___ ___   ___ ___ ___ ___ ___ ___

9.   1st term 30, step 6.   1st term 34, step 8.   1st term 42, step 12.
    What's the 4th term?   What's the 5th term?   What's the 6th term?

    ___ ___ ___ ___   ___ ___ ___ ___ ___   ___ ___ ___ ___ ___ ___

Ch 12 Ls 2 Arithmetic Sequence Story Problems

_____ #1 #2 ____/ 7 #3 ____/ 6 R ____/ 7 Total ____/ 20 _____
Name   Checker

#1 1. What is the sequence formula? _____

2. What does a story problem need to use a sequence? _____
_____

3. Name 3 story problems that use sequence. _____
_____

#2 1. Make an equation. What starts it?   Find the 7th step in a sequence that starts at 1 with steps of 2.

What adds to the 1st term?   _____

What multiplies with it?   _____

What is the 7th step?   _____

                             _____

2. What is happening? $4 + (8 - 1)5$

What are the numbers?   Start at ____ with ____ steps of ____.

                             __ __ __ __ __ __ __ __

3. What is happening? $9 + (4 - 1)10$

What are the numbers?   Start at ____ with ____ steps of ____.

                             __ __ __ __

4. What is happening? $8 + (7 - 1)6$

What are the numbers?   Start at ____ with ____ steps of ____.

                             __ __ __ __ __ __ __

#3 1. 5 + (5 - 1)4   8 + (6 - 1)7   Calculator? yes no

Find the Start at ___ with ___ steps of ___.   Start at ___ with ___ steps of ___.
numbers. ___ ___ ___ ___ ___   ___ ___ ___ ___ ___ ___

2. 2 + (5 - 1)6   10 + (7 - 1)10

Start at ___ with ___ steps of ___.   Start at ___ with ___ steps of ___.

___ ___ ___ ___ ___   ___ ___ ___ ___ ___ ___ ___

3. 3 + (4 - 1)8   12 + (6 - 1)5

Start at ___ with ___ steps of ___.   Start at ___ with ___ steps of ___.

___ ___ ___ ___ ___   ___ ___ ___ ___ ___ ___ ___

Review 1. What is the sequence formula? _____   Calculator?
yes no
2. What does a story problem need to use a sequence? _____

_____

3. Name 3 story problems that use sequence. _____

_____

Story problems.

4. The chipmunk population in a garden is growing at an alarming rate. The counts taken show there were 12 frogs to start, then 18, then 24. If they continue to grow at this rate, what will the next count be?   _____ chimpmunks

5. Zara is on a road trip. She has 445 k to go, then 367 k an hour later and 299 k an hour after that. What's next?   _____ kilometers

6. James borrowed Rs 915 from the bank in January. In February he owed Rs 837. In March he owed Rs 759. What will it be in April?   _____ rupees

7. Ojas is a biologist. He counted 48 bacteria on day 1, 120 on day 2, 192 on day 3, and 264 on day 4. If the bacteria continue at this rate, what will day 5 show?   _____ bacteria

Ch 12 Ls 3 Arithmetic Series Formula   253

_____ #1 #2 ____ / 8 #3 ____ / 6 R ____ / 9 T ____ / 23 _____
        Name   Checker

#1 1. What does the arithmetic series do? _____
    2. Look at the terms. What is it adding? $(a_1 + a_n)$ _____
    3. What is it multiplied with? $N(a_1 + a_n)$ _____
    4. Divide by 2. What does that show you? _____
    5. What is happening in this equation? $\dfrac{4(3 + 9)}{2}$ _____

#2 1. What's inside the parentheses? What does it multiply?   **1, 4, 7, 10**

                    What does it divide by?   ___(+)

                    Solve it. What's the answer?   ___

                                                   _____

    2. What's inside the parentheses? What does it multiply?   **3, 10, 17, 24**

                    What does it divide by?   ___(+)

                    Solve it. What's the answer?   ___

                                                   _____

    3. What's inside the parentheses? What does it multiply?   **5, 13, 21, 29, 37**

                    What does it divide by?   ___(+)

                    Solve it. What's the answer?   ___

                                                   _____

#3 Make the equation, then solve it.   Calculator?
                                              yes   no

1. **1, 5, 9, 13**   **4, 8, 12, 16, 20**   **3, 5, 7, 9, 11, 13**
    ___(+)   ___(+)   ___(+)
    ___   ___   ___

    _____   _____   _____

2. **8, 15, 22, 29**   **6, 9, 12, 15, 18**   **8, 14, 20, 26, 33**
    ___(+)   ___(+)   ___(+)
    ___   ___   ___

    _____   _____   _____

Review 1. What does the arithmetic series do? _____   Calculator?
        2. Look at the terms. What is it adding? $(a_1 + a_n)$ _____   yes   no
        3. What is it multiplied with? $N(a_1 + a_n)$ _____
        4. Divide by 2. What does that show you? _____
        5. What is happening in this equation? $\frac{4(3 + 9)}{2}$ _____

        What is happening in each expression? Solve it.

6. $\frac{5(3 + 11)}{2}$ $\frac{6(5 + 20)}{2}$

    Start: ___ End: ___. ___ steps of ___.   Start: ___ End: ___. ___ steps of ___.

    _____   _____

7. $\frac{4(7 + 25)}{2}$ $\frac{7(4 + 28)}{2}$

    Start: ___ End: ___. ___ steps of ___.   Start: ___ End: ___. ___ steps of ___.

    _____   _____

Ch 12 Ls 4 Arithmetic Sigma/Story Problems   255

_____ #1 #2 ____ / 6 #3 ____ / 4 R ____ / 5 Total ____ / 15 _____
        Name   Checker

#1 1. How do you say this sigma problem? $\sum_{n=1}^{4} 3n$ _____

2. Name 3 steps to solve sigma problems. _____
_____

3. What does an arithmetic series story problem need? _____

#2 1. Solve as a series. Find the 1st step. $\sum_{n=2}^{6} 2n + 2$

Find the last step. _____

Make an equation. What starts it? _____

What does it divide? Solve it. _____

2. Solve as a series. What are the 1st and last numbers? $\sum_{n=3}^{7} 4n + 1$

Make an equation. What starts it? _____

What does it divide? Solve it. _____

3. Make an equation.   During Kwanzaa a candle a candle is lit, then the next day 2 are lit, until 7 are lit. How many are lit during the 7 day festival.

Solve a step. _____

What's the answer? _____

1. $\sum_{n=1}^{4} 4n + 1$ 	 $\sum_{n=1}^{4} 5n - 4$ 	 Calculator? yes no

   ___(___ + ___) = ___ 	 ___(___ + ___) = ___

   ___ ÷ 2 = ___ 	 ___ ÷ 2 = ___

2. $\sum_{n=1}^{4} 7n - 3$ 	 $\sum_{n=1}^{4} 8n + 5$

   ___(___ + ___) = ___ 	 ___(___ + ___) = ___

   ___ ÷ 2 = ___ 	 ___ ÷ 2 = ___

Review 1. How do you say this sigma problem? $\sum_{n=1}^{4} 3n$ _____ Calculator? yes   no

2. Name 3 steps to solve sigma problems. _____

_____

3. What does an arithmetic series story problem need? _____

4. What's the equation?   A construction company is fined Rs 200 the 1st day and doubled each day until work is complete. They finish 5 days late, so they pay all of the fines. How much was their fine?

   What's the 1st step?   _____

   What's the answer?   _____

   _____

5. What's the equation?   A shopping company pays Rs 100 the first day, Rs 200 the next day, and Rs 400 the 3rd day until their work is finished. They pay all the fines. How much did they owe?

   What's the 1st step?   _____

   What's the answer?   _____

   _____

Review Problems 257

_____ #1 - #3 ____/10 #4 ____/ 7 Total ____/ 17
            Name

1. **Arithmetic Sequence** _____
2. **Arithmatic Series** _____

#2 Decide the step. Find the missing term.   Calculator?
                                                               yes   no
#2 1.   2 + (8 - 1)3   6 + (7 - 1)7

           Start at ___ with ___ steps of ___.   Start at ___ with ___ steps of ___.
Solve
them.   _____   _____

    2.   4 + (9 - 1)6   8 + (5 - 1)10

           Start at ___ with ___ steps of ___.   Start at ___ with ___ steps of ___.

           _____   _____

    3.   7 + (6 - 1)7   10 + (7 - 1)5

           Start at ___ with ___ steps of ___.   Start at ___ with ___ steps of ___.

           _____   _____

                    #3 Make the equation, then solve it.   Calculator?
    1. 1, 6, 11, 16   4, 7, 10, 13, 16   3, 7, 11, 15, 19   yes   no

        ___(+)   ___(+)   ___(+)
        ___   ___   ___

        _____   _____   _____

    2. 7, 15, 23, 31   5, 9, 13, 17, 21   10, 15, 20, 25, 30

        ___(+)   ___(+)   ___(+)
        ___   ___   ___

        _____   _____   _____

258

#3

Solve these story problems.

Calculator? yes no

1. A radio station has a contest where they give away Rs 1000 and each day they add Rs 95, the station's call number. You win it on the 9th day. How much did you win?

2. The floor on a bathroom starts with 1 tile, then 2 tiles and on until 8th row is laid. How many tiles would you buy for the floor?

3. Ojas swims 1.0 km on Monday, 1.1 km on Tuesday, 1.3 km on Wednesday, 1.6 km on Thursday, and 2.0 km on Friday. If the pattern continues, how many kilometers will he swim on Saturday?

4. A policeman donates the same amount of money each year to a land fund. Each year he's donated enough money to protect 5 hectoacres. How many acres will the policeman's donations protect at the end of the 12th year?

5. An embroidery pattern calls for four stitches in the first row and for three more stitches in each successive row. The 25th row, which is the last row, has 76 stitches. Find the total number of stitches in the pattern.

6. Each year, a volunteer organization expects to add 7 more people to the number of shut ins. This year, the organization provides the service for 42 people. How many years have they been in business?

7. This month, your friend deposits Rs 4000 to save for a vacation. She plans to deposit 10% more each successive month for the next 13 months. How much will she have saved after the 14 deposits?

Ch 13 Ls 1 Geometric Sequence Formula. 259

_____ #1 #2 ____ / 8 #3 ____ /10 R ____ / 5 Total ____ / 23 _____
         Name   Checker

#1 1. What 2 things does a geometric sequence multiply? _____
       2. What is the ratio? _____
       3. What makes the exponent? _____
       4. What is the formula for a geometric sequence? _____
       5. How do you remember it? _____

#2 1. What is happening in this problem?   2 4 8 16 32

              What's the equation?   First term: ___ Ratio: ___ Steps: ___

              Solve it. What's the answer?   ___ X ___ ―

                                                   = ___

       2. What is happening in this problem?   5 10 20 40

              What's the equation?   First term: ___ Ratio: ___ Steps: ___

              Solve it. What's the answer?   ___ X ___ ―

                                                   = ___

       3. What starts the equation?   **Find the first 3 terms in a geometric sequence that starts with 4 and has ratio of 2.**

              What's the exponent?   _____

              What does it multiply?   _____

              Solve it. What's the answer?   _____

                                                _____

#3 1. $3 \times 2^{6-1}$ $5 \times 3^{5-1}$ Calculator? yes no

First term: ___ Ratio: ___ Steps: ___   First term: ___ Ratio: ___ Steps: ___

What is happening in each problem? Then find the answer.

$3 \times 2 \longrightarrow =$ ___ $5 \times 3 \longrightarrow =$ ___

2. $4 \times 5^{4-1}$ $2 \times 2^{7-1}$

First term: ___ Ratio: ___ Steps: ___   First term: ___ Ratio: ___ Steps: ___

$4 \times 5 \longrightarrow =$ ___ $2 \times 2 \longrightarrow =$ ___

3. $8 \times 2^{6-1}$ $6 \times 3^{5-1}$

First term: ___ Ratio: ___ Steps: ___   First term: ___ Ratio: ___ Steps: ___

$8 \times 2 \longrightarrow =$ ___ $6 \times 3 \longrightarrow =$ ___

4. $4 \times 5^{3-1}$ $7 \times 3^{4-1}$

First term: ___ Ratio: ___ Steps: ___   First term: ___ Ratio: ___ Steps: ___

$4 \times 5 \longrightarrow =$ ___ $7 \times 3 \longrightarrow =$ ___

5. $5 \times 4^{5-1}$ $10 \times 2^{4-1}$

First term: ___ Ratio: ___ Steps: ___   First term: ___ Ratio: ___ Steps: ___

$5 \times 4 \longrightarrow =$ ___ $10 \times 2 \longrightarrow =$ ___

Review 1. What 2 things does a geometric sequence multiply? _____   Calculator? yes no

2. What is the ratio? _____

3. What makes the exponent? _____

4. What is the formula for a geometric sequence? _____

5. How do you remember it? _____

Ch 13 Ls 2 Geometric Sequence Story Problems   261

_____ #1 #2 ____/ 6   #3 ____/ 4   R ____/ 3   Total ____/ 13 _____
        Name   Checker

#1 1. How can you tell a negative ratio? _____
       2. How can you tell a fraction ratio? _____
       3. What kind of story problem uses geometric sequence? _____
       _____

#2 1. What is happening in this problem?   5 10 20 40

          What's the equation?   First term: ___ Ratio: ___ Steps: ___

          Solve it. What's the answer?   ___ X ___ —

                                               = ___

       2. What is happening in this problem?   2 8 32 128

          What's the equation?   First term: ___ Ratio: ___ Steps: ___

          Solve it. What's the answer?   ___ X ___ —

                                               = ___

       3. What's the   JJ has purchased 7 books. The 1st book cost R 1, the 2nd,
          equation?   Rs 2, the 3rd, Rs 4, and the 4th, Rs 8, and so on. How
                         expensive was the 7th book?

          What's the 1st step?   _____

          What's the answer?   _____

                                 _____

#3 Find the equation and solve each problem.

Calculator? yes no

1. Mr. K suffers from allergies. When allergy season arrives, his doctor recommends that he take 500 mg of his medication the first day, and decrease the dosage by one half each day for 4 days. Which "rule" represents his medication doses for the week?

____ X ____
_____
_____
_____

2. Ira has decided to add strength training to her exercise program. Her father suggests that she add weight lifting for 5 minutes during her program for the first week. Each week thereafter, she is to increase the weight lifting time by 2 times. How long will her strength training be in 4 weeks?

____ X ____
_____
_____
_____

3. A laboratory is to begin experimentation with a bacteria that doubles every 4 hours. The lab starts with 100 bacteria. How many bacteria would there be after 24 hours?

____ X ____
_____
_____
_____

4. A prize starts at Rs 400 the 1st day and adds a ratio of x 2 starting with Rs 40 each day. You win on the 8th day. How much did you win?

____ X ____
_____
_____
_____

Review 1. How can you tell a negative ratio? _____

Calculator? yes no

2. How can you tell a fraction ratio? _____

3. What kind of story problem uses geometric sequence? _____

_____

Ch 13 Ls 3 Geometric Series Formula 263

_____ #1 #2 ____/ 8 #3 ____/ 6   R ___/ 9 Total ____/ 23 _____
           Name   Checker

#1 1. What starts the geometric series formula? _____

2. What fraction does it multiply? _____

3. What do the 3 variables stand for? _____
_____

4. What are the 1st 2 steps to solve it? _____

5. How do you say a geometric series? _____

#2 1. What is the fraction part?   **1, 3, 9, 27**

What does it multiply? $\dfrac{1\ -\ }{1\ -\ }$ ___

Solve it.
What's the answer?   ___ $\dfrac{1\ -\ }{1\ -\ }$ ___

_____

2. What is the fraction part?   **1, 4, 7, 10**

What does it multiply? $\dfrac{1\ -\ }{1\ -\ }$ ___

Solve it.
What's the answer?   ___ $\dfrac{1\ -\ }{1\ -\ }$ ___

_____

3. What does it multiply?   **2, 6, 18, 54**

Solve it.
What's the answer?   ___ $\dfrac{1\ -\ }{1\ -\ }$ ___

_____

#3 1.   **1, 3, 7**   **2, 4, 8**   **2, 6, 14**   Calculator?
                                                                    yes   no

Find the equation
and solve it.

2. **1, 5, 13, 25**   **5 10 20 40**   **3, 6, 12, 24**

                                                                    Calculator?
Review 1. What starts the geometric series formula? _____   yes   no

2. What fraction does it multiply? _____

3. What do the 3 variables stand for? _____
_____

4. What are the 1st 2 steps to solve it? _____

5. How do you say a geometric series? _____

6. $1 \cdot \dfrac{1-3^3}{1-3}$ $1 \cdot \dfrac{1-5^3}{1-5}$

What is happening?
Then solve it.

1st term: ___ Ratio: ___ # terms: ___   1st term: ___ Ratio: ___ # terms: ___

_____   _____

7. $1 \cdot \dfrac{1-4^3}{1-4}$ $1 \cdot \dfrac{1-7^3}{1-7}$

1st term: ___ Ratio: ___ # terms: ___   1st term: ___ Ratio: ___ # terms: ___

_____   _____

Ch 13 Ls 4 Geometric Series Story Problem   265

_____ #1 #2 ____/ 5 #3 ____/ 6 R ____/ 4 Total ____/ 15 _____
              Name   Checker

#1 1. What does geometric series do? _____

2. What does an geometric series story problem need? _____

_____

#2 1. What starts   A prize starts at Rs 200 and adds Rs 50, then Rs 100, and Rs 200   Equation?
       the equation? each day until it's won. You win on the 7th day. What's the prize?

What adds to the 1st term?   _____

What multiplies next?   _____

What is the 10th step?   _____

                              _____

2. What's the   Another prize starts at Rs 100 and adds Rs 70, then Rs 140,   Equation?
   equation?   and Rs 280 each day. You win on the 5th day. How much is it?

                              _____

What's the answer?   _____

                              _____

3. What's the   JJ has purchased 7 books. The 1st book cost R 1, the 2nd, Rs
   equation?   2, the 3rd, Rs 4, and so on. How much did all 7 books cost? Equation?

                              _____

What's the 1st step?   _____

What's the answer?   _____

                              _____

1.
What is happening?
Then solve it.

$1 \cdot \dfrac{1-5^3}{1-5}$ $2 \cdot \dfrac{1-3^4}{1-3}$ Calculator? yes   no

1st term: ___ Ratio: ___ # terms: ___   1st term: ___ Ratio: ___ # terms: ___

_____   _____

2. $5 \cdot \dfrac{1-6^3}{1-6}$ $10 \cdot \dfrac{1-5^3}{1-5}$

1st term: ___ Ratio: ___ # terms: ___   1st term: ___ Ratio: ___ # terms: ___

_____   _____

3. $7 \cdot \dfrac{1-4^3}{1-4}$ $8 \cdot \dfrac{1-7^4}{1-7}$

1st term: ___ Ratio: ___ # terms: ___   1st term: ___ Ratio: ___ # terms: ___

_____   _____

Review 1. What does geometric series do? _____   Calculator? yes no

2. What does an geometric series story problem need? _____

Make and solve this equation.

3. Your father wants you to help him build a shed in the backyard. He says he will pay you $5 for the first week and then double it every week. The project will take 6 weeks. How much money will you earn, in total, if you work for the 6 weeks?

_____
_____

4. Your grandmfaher gives you $100 to start a college fund. He then tells you he'll add a ratio of 2 starting with $5 for 10 months or you can just take $1000. Which one should you take?

_____
_____
_____

Review Problems 267

_____ #1 - #3 ____/ 13 #4 ____/ 6 Total ____/ 19
          Name

1. Geometric Sequence _____
2. Geometric Series _____
3. infinite series _____

#2   1. $2 \times 3^{4-1}$ $4 \times 3^{5-1}$ Calculator?
                                                                          yes   no
         First term: ___ Ratio: ___ Steps: ___   First term: ___ Ratio: ___ Steps: ___

What is happening in
 each problem? $2 \times 3 \text{—} = $ ___ $4 \times 3 \text{—} = $ ___
Then find the answer.

         2. $5 \times 5^{4-1}$ $3 \times 2^{7-1}$

         First term: ___ Ratio: ___ Steps: ___   First term: ___ Ratio: ___ Steps: ___

                 $5 \times 5 \text{—} = $ ___ $3 \times 2 \text{—} = $ ___

         3. $7 \times 2^{6-1}$ $9 \times 3^{5-1}$

         First term: ___ Ratio: ___ Steps: ___   First term: ___ Ratio: ___ Steps: ___

                 $7 \times 2 \text{—} = $ ___ $9 \times 3 \text{—} = $ ___

#3   1. $1 \dfrac{1-4^3}{1-4}$ $3 \dfrac{1-3^4}{1-3}$ Calculator?
What is happening?   yes   no
 Then solve it.

         1st term: ___ Ratio: ___ # terms: ___   1st term: ___ Ratio: ___ # terms: ___

         _____   _____

         2. $5 \dfrac{1-7^3}{1-7}$ $8 \dfrac{1-5^3}{1-5}$

         1st term: ___ Ratio: ___ # terms: ___   1st term: ___ Ratio: ___ # terms: ___

         _____   _____

                                                  Finish it on the next page.

268

3. $\quad 7\,\dfrac{1-6^3}{1-6} \qquad\qquad 10\,\dfrac{1-5^3}{1-5}$ Calculator?
yes   no

1st term: ___ Ratio: ___ # terms: ___   1st term: ___ Ratio: ___ # terms: ___

_____   _____

#4 Find the equation and solve them.

1. Mr. K suffers from allergies. When allergy season arrives, his doctor recommends that he take 500 mg of his medication the first day, and decrease the dosage by one half each day for 4 days. Which "rule" represents his medication doses for the week?

   _____
   _____
   _____
   _____

   Calculator?
   yes   no

2. Ira has decided to add strength training to her exercise program. Her doctor suggests that she add weight lifting for 5 minutes during her program for the first week. Each week thereafter, she is to increase the weight lifting time by 2 times. How long will her strength training be in 4 weeks?

   _____
   _____
   _____
   _____

3. Mr. K suffers from allergies. When allergy season arrives, his doctor recommends that he take 900 mg of his medication the first day, and decrease the dosage by one half each day for 4 days. Make an equation and solve for the final dose.

   _____
   _____
   _____
   _____
   _____

4. A runner begins training by running 4 km. one week. The second week she runs a total of 6 km. The third week she runs 10 km. Assume this pattern continues. How far will she run in the 6th week? Make an equation and solve it

   _____
   _____
   _____
   _____

Ch 14 Ls 1 Infinite Series Formula 269

_____ #1 #2 ____/ 8 #3 ____/ 9 R ____/ 5 Total ____/ 22 _____
        Name   Checker

#1 1. What is the formula for an infinite series? _____
        2. What does R have to be to converge? _____
        3. What is the key question for infinite series? _____
        4. What makes it an infinite series? _____
        5. What does r have to be to converge it? _____

#2 1. Solve these infinite series problems.

$$\frac{20}{1 - \frac{1}{2}} \qquad \frac{20}{1 - 3}$$ What's first?

_____   _____

_____   _____

2. Make an equation. What's the fraction?   **First term is 10 with ratio of 1 4th.**

   Solve the 1st step.   _____

   What's the answer?   _____

                                            _____

3. Make an equation. What's the fraction?   **First term is 14 with ratio of 5.**

   Solve the 1st step.   _____

   What's the answer?   _____

                                            _____

#3 Make an equation and solve, then find what's happening.   Calculator?
                                                                    yes   no

1. **1st term 20, ratio is 1 4th.**   **1st term 10, ratio is 4.**   **1st term 6, ratio is 3 4ths.**

$$\frac{\quad\quad}{1-\quad}\qquad \frac{\quad\quad}{1-\quad}\qquad \frac{\quad\quad}{1-\quad}$$

_____   _____   _____
_____   _____   _____

2. **1st term 30, ratio is 2 5ths.**   **1st term 40, ratio is 1/10.**   **1st term 50, ratio is 5.**

$$\frac{\quad\quad}{1-\quad}\qquad \frac{\quad\quad}{1-\quad}\qquad \frac{\quad\quad}{1-\quad}$$

Start ___ Ratio is ___.   Start at ___ Ratio is ___.   Start at ___ Ratio is ___.

_____   _____   _____
_____   _____   _____

3. **1st term 10, ratio is 2.**   **1st term 14, ratio is 0.80.**   **1st term 25, ratio is 1 5th.**

$$\frac{\quad\quad}{1-\quad}\qquad \frac{\quad\quad}{1-\quad}\qquad \frac{\quad\quad}{1-\quad}$$

Start ___ Ratio is ___.   Start at ___ Ratio is ___.   Start at ___ Ratio is ___.

_____   _____   _____
_____   _____   _____

Review 1. What is the formula for an infinite series? _____   Calculator?
        2. What does R have to be to converge? _____   yes   no
        3. What is the key question for infinite series? _____
        4. What makes it an infinite series? _____
        5. What does r have to be to converge it? _____

Ch 14 Ls 2 Sigma and Infinite Series 271

_____ #1 #2 ____ / 8 #3 ____ / 6 R ____ / 9 Total ____ / 23 _____
Name   Checker

$$\sum_{n=1}^{\infty} cr^{n-1}$$

#1 1. How do you know this is an infinite series problem? _____

2. Where is the first term? _____

3. What does r stand for? _____

4. What does n stand for? _____

#2 1. Make an equation. What starts it? $\sum_{n=1}^{\infty} 5\left(-\frac{1}{2}\right)^{n-1}$

Solve the 1st step.   _____

What's the answer?

2. Make an equation. What starts it? $\sum_{n=1}^{\infty} 10\left(\frac{1}{2}\right)^{n-1}$

Solve the 1st step.   _____

What's the answer?

3. Write the equation. $\sum_{n=1}^{\infty} 25\left(-\frac{1}{4}\right)^{n-1}$ $\sum_{n=1}^{\infty} 35\left(\frac{1}{4}\right)^{n-1}$
   Then solve it.

#3 1. $\sum_{n=1}^{\infty} 35 \left(\frac{1}{4}\right)^{n-1}$ $\sum_{n=1}^{\infty} 35 \left(\frac{1}{4}\right)^{n-1}$ Calculator? yes no

_____   _____

_____   _____

_____   _____

2. $\sum_{n=1}^{\infty} 35 \left(\frac{1}{4}\right)^{n-1}$ $\sum_{n=1}^{\infty} 35 \left(\frac{1}{4}\right)^{n-1}$

_____   _____

_____   _____

_____   _____

$\sum_{n=1}^{\infty} cr^{n-1}$

Review 1. How do you know this is an infinite series problem? _____   Calculator?
2. Where is the first term? _____   yes no
3. What does r stand for? _____
4. What does n stand for? _____
5. **A girl on a swing is given a push. She travels 12 meters on the first back-and-forth swing but only 3/4 as far on each successive back-and-forth swing. How far (total distance) does she travel before the swing stops?**

   _____

   _____

   _____

6. **A ball is thrown 14 meters in the air (so that the up-and-down distance is 28 meters). The ball rebounds 90% of the distance it falls. What is the total vertical distance traveled by the ball before it stops bouncing?**

   _____

   _____

Ch 14 Ls 3 Fibonacci Sequence/Fractal Geometry   273

_____ #1 #2 ____ / 6 #3 ____ / 4 R ____ / 4 Total ____ / 14 _____
           Name   Checker

#1 1. What is the Fibonacci sequence? 1, 1, 2_____

2. What's the 1st step for Fibonacci? $f(x) = 3x + 2$ for $x_0 = 1$_____

_____

3. What's the 1st step for Fibonacci? _____

4. What starts the equation?   A prize starts at Rs 10 and adds Rs 20 to it. The next day it adds Rs 30 to it to make it Rs 50. The next day it adds Rs 40, so it's Rs 70. You win on the 8th day. How much did you win?

What adds to the 1st term?   _____

What multiplies next?   _____

What is the 8th step?   _____

                                    _____

#2 1. What is Fractal Geometry? _____

2. What is the next step with this Fractal? ─── ⋀ _____

_____

3. What is the next step with this Fractal? | _____

4. What's the 2nd step with it? _____

5. What's the 1st part to this triangle having Fractals? ▼ _____

_____

6. What's the 2nd part to this triangle having Fractals? _____

#3 Fibonacci Word Problems

Calculator?
yes   no

A teacher has a student led awards program where the class receives 5 minute free time, then 6 minutes the next day, and 11 minutes the 3rd day. On the 5th day how many minutes free time will they get?

1. How many will they get 4th day? _____

2. How many will they get 5th day? _____

Ojas starts off his new job without much money, so he put R 1 in his cup for savings. He started making more money, so he put Rs 2, then Rs 3 and Rs 5 in his cup. He kept on adding the last 2 weeks and put that in his cup.

3. How many will they get 4th pay? _____

4. How many will they get 7th pay? _____

5. How many will they get 10th pay? _____

6. How many will they get 12th pay? _____

Review 1. What is the Fibonacci sequence? **1, 1, 2** _____

2. What's the 1st step for Fibonacci? $f(x) = 3x + 2$ for $x_0 = 1$ _____

Calculator?
yes   no

_____

3. What's the 1st step for Fibonacci? _____

4. What is Fractal Geometry? _____

5. What is the next step with this Fractal? ___ ⋀ _____

_____

6. What is the next step with this Fractal? | _____

7. What's the 2nd step with it? _____

8. What's the 1st part to this triangle having Fractals? ▼ _____

_____

9. What's the 2nd part to this triangle having Fractals? _____

Review Problems **275**

_____ #1 - #3 ____ / 11 #3 #4 ____ / 16 Total ____ / 27
Name

1. infinite series _____

2. Fibonacci Sequence _____

3. Fractal Geometry _____

#2 1. 1st term 20, ratio is 1 halveth.   1st term 10, ratio is 1 4th.   1st term 10, ratio is 1 4th. Calculator?
                                                                                                                                                        yes   no

$$\frac{20}{1 - \frac{1}{2}} \qquad \frac{10}{1 - \frac{1}{4}} \qquad \frac{30}{1 - \frac{1}{5}}$$

_____   _____   _____

_____   _____   _____

2. 1st term 40, ratio is 3 4ths.   1st term 10, ratio is 2 5ths.   1st term 50, ratio is 3 5ths.

$$\frac{40}{1 - \frac{3}{4}} \qquad \frac{14}{1 - \frac{2}{5}} \qquad \frac{50}{1 - \frac{3}{5}}$$

Start ___ Ratio is ___ .   Start at ___ Ratio is ___ .   Start at ___ Ratio is ___ .

_____   _____   _____

_____   _____   _____

#3. 1. $$\sum_{n=1}^{\infty} 35 \left(\frac{1}{4}\right)^{n-1} \qquad \sum_{n=1}^{\infty} 16 \left(\frac{1}{2}\right)^{n-1}$$ Calculator?
                                                                                                                                                        yes   no

_____   _____

_____   _____

_____   _____

_____   _____

Continued on the next page....

2. $\sum_{n=1}^{\infty} 24 \left(\frac{3}{4}\right)^{n-1}$ $\sum_{n=1}^{\infty} 32 \left(\frac{7}{8}\right)^{n-1}$

_____

_____

_____

3. $\sum_{n=1}^{\infty} 42 \left(\frac{1}{2}\right)^{n-1}$ $\sum_{n=1}^{\infty} 45 \left(\frac{3}{5}\right)^{n-1}$

_____

_____

_____

4. $\sum_{n=1}^{\infty} 56 \left(\frac{1}{4}\right)^{n-1}$ $\sum_{n=1}^{\infty} 60 \left(\frac{4}{5}\right)^{n-1}$

_____

_____

_____

#4

What is the Fibonacci Sequence?

1. 2, 6, 8, ____, ____   3, 5, 8, ____, ____   Calculator? yes   no
2. 4, 10, 14, ____, ____   5, 6, 11, ____, ____
3. 6, 7, 13, ____, ____   8, 9, 17, ____, ____
4. 2, 5, 7, ____, ____   14, 16, 30, ____, ____
5. 20, 25, 45, ____, ____   35, 45, 80, ____, ____

Ch 1 Ls 1 Basic Exponent Form 277

_____ #1 #2 ____ /31 #3 ____ /18 R ____ /28 Total ____ /77 _____
      Name   Checker

#1 1. What are the 3 parts to exponent form? _____
   2. What do logarithms find as "the number"? _____
   3. How do exponents use the Rule of 3s? _____
   4. What does the exponent 2 on base 10 show you? 10^2 _____

#2 1. What are 10 to the 4th and 8th? 10^4 10^8

                                          _____   _____

   2. What are 10 to the 6th and 3rd? 10^6 10^3

                                          _____   _____

   3. What exponent finds 10 million, base 10? $10^? = 10{,}000{,}000$

                                                        _____

   4. What exponent finds 1 billion, base 10? $10^? = 1{,}000{,}000{,}000$

                                                        _____

   5. Find the base 2s. 2^2 2^3 2^4 2^5 2^6 2^7 2^8

                            ___   ___   ___   ___   ___   ___   ___

   6. Find the base for 3^2 3^3 3^4 3^5 4^2 4^3 4^4
      3s and 4s.
                            ___   ___   ___   ___   ___   ___   ___

   7. Find a base 5s, 5^2 5^3 5^4 6^2 6^3 7^2 7^3
      6s and 7s.
                            ___   ___   ___   ___   ___   ___   ___

#3 Solve these Exponents.   Calculator?
                                               yes   no

1. $10^6 =$ _____ $10^4 =$ _____ $10^8 =$ _____

2. $10^3 =$ _____ $10^2 =$ _____ $10^5 =$ _____

3. $10^9 =$ _____ $10^7 =$ _____ $10^1 =$ _____

4. $10^- = 1,000$ $10^- = 1,000,000$ $10^- = 100,000$

5. $10^- = 10,000$ $10^- = 10,000,000$ $10^- = 10$

6. $10^- = 100$ $10^- = 10,000,000,000$ $10^- = 100,000,000$

Review 1. What are the 3 parts to exponent form? _____   Calculator?
                                                                            yes   no
2. What do logarithms find as "the number"? _____
3. How do exponents use the Rule of 3s? _____
4. What does the exponent 2 on base 10 show you? **10** _____

5. $3^- = 27$ $2^- = 8$ $4^- = 16$ $5^- = 25$

6. $2^- = 16$ $6^- = 36$ $2^- = 128$ $3^- = 9$

7. $4^- = 64$ $2^- = 4$ $7^- = 49$ $6^- = 216$

8. $2^- = 32$ $5^- = 625$ $4^- = 256$ $8^- = 81$

9. $3^- = 81$ $2^- = 64$ $8^- = 648$ $10^- = 100$

10. $9^- = 81$ $10^- = 1,000$ $5^- = 125$ $2^- = 256$

                                        Each lesson has a quiz.

Ch 1 Ls 2 Change Logarithm to Exponent Form 279

_____ #1 #2 ____/10 #3 #4 ____/ 18 R ____/ 8 Total ____/ 36 _____
Name   Checker

#1 1. What parts are on the left side of a logarithm? _____
      2. What does a logarithm equal? _____
      3. Where do you start to say a logarithm? _____
      4. What sign shows how to say a logarithm? _____

        5. Where do you start to say a logarithm? $\log_{10} 100 = 2$

              Where does it go to next?   _____

              Where does the log go next?   _____

                                                 _____

#2 1. What's the log exponent for 100,000? $\log 100{,}000 = ?$

                                                  _____

      2. What's the log exponent for 10 million? $\log 10{,}000{,}000 = ?$

                                                  _____

      3. What real number has 3 places? $\log_{10} ? = 3$

                                                  _____

      4. What real number has 4 places? $\log_{10} ? = 4$

                                                  _____

      5. What real number has 6 places? $\log_{10} ? = 6$

                                                  _____

#3 Find the log exponent.   Calculator?
                                                yes   no

1.   log 1000 = ___   log 100,000 = ___

2.   log 1,000,000 = ___   log 10,000 = ___

3.   log 10,000,000 = ___   log 1 = ___

4.   log 100 = ___   log 10,000,000,000 = ___

#4 Find the real number.   Calculator?
                                                yes   no

1. log _____ = 2   log _____ = 5

2. log _____ = 4   log _____ = 6

3. log _____ = 7   log _____ = 1

4. log _____ = 10   log _____ = 9

5. log _____ = 3   log _____ = 8

Review 1. What parts are on the left side of a logarithm? _____   Calculator?
        2. What does a logarithm equal? _____   yes   no
        3. Where do you start to say a logarithm? _____
        4. What sign shows how to say a logarithm? _____
        5. Change the log exponent to a 3, what does it change? $\log_{10} x = 3$ _____
        _____

Write how to say each log
6. log 100,000 = 5   _____
7. log 100 = 2   _____
8. log 10,000 = 4   _____

Ch 1 Ls 3 Change Exponent to Logarithm Form   281

_____ #1 #2 ____/ 8 #3 #4 ____/20 R ___/ 3 Total ____/ 31 _____
          Name   Checker

#1 1. What letter shows how to change exponent to log form? $10^2 = 100$ _____

2. Where do you start to change an exponent to log form? _____

_____

3. Where does it go next after the real number? _____

_____

#2 1. Where does a log start and go to? $10^3 = 1000$

         Start at log _____ base _____   Finish it.

         Start at log _____ base _____ is _____

2. Where does a log start and go to? $10^4 = 10{,}000$

         Start at log _____ base _____   Finish it.

         Start at log _____ base _____ is _____

3. Where does a log start and go to? $10^6 = 1{,}000{,}000$

         Start at log _____ base _____ is _____

4. What's the log form for 10 squared? $10^2 = 100$

         $\log_{10}$ _____ = _____

5. What's the log form for 10 to the 5th? $10^5 = 100{,}000$

         $\log_{10}$ _____ = _____

#3 Change these exponents to log form.

1. 10^7 is log _____ = ___ 10^3 is log _____ = ___   Calculator?
                                                                                                                       yes   no

2. 10^5 is log _____ = ___ 10^9 is log _____ = ___

3. 10^6 is log _____ = ___ 10^4 is log _____ = ___

4. 10^2 is log _____ = ___ 10^8 is log _____ = ___

5. 10^1 is log _____ = ___ 10^{10} is log _____ = ___

#4 Change these logarithms to exponent form.   Calculator?
                                                                                                                                      yes   no

1.   log 10 = 1 $10^{__} = $ ___   log 100,000,000 = 8 $10^{__} = $ ___

2.   log 10,000 = 4 $10^{__} = $ ___   log 100 = 2 $10^{__} = $ ___

3.   log 100,000 = 5 $10^{__} = $ ___   log 10,000,000 = 7 $10^{__} = $ ___

4.   log 1,000 = 3 $10^{__} = $ ___   log 10,000,000,000 = 10 $10^{__} = $ ___

5.   log 1,000,000 = 6 $10^{__} = $ ___   log 1,000,000,000 = 9 $10^{__} = $ ___

Review 1. What letter shows how to change exponent to log form? $10^2 = 100$ _____   Calculator?
                                                                                                                                                   yes   no

_____

2. Where do you start to change an exponent to log form? _____

_____

3. Where does it go next after the real number? _____

Ch 2 Ls 1 Find Log Decimals for 2, 4, 3, and 6   283

_____ #1 #2 ____/10 #3 #4 ____/16 R ___/14 Total ____/ 40 _____
            Name   Checker

#1 1. Name the log of 2. How can you remember it? _____

2. How can you remember the log of 4? _____

3. Name 2 parts to a log exponent. **Log 20 = 1.30** _____

_____

4. Find the exponent. What's the place value?   **log 20 = ?**

Place Value is ____   **Finish it.**

log 20 = ____

5. Find the exponent. What's the place value?   **log 400 = ?**

Place Value is ____   **Finish it.**

log 400 = ____

#2 1. What is the log exponent for 3? _____

2. What is the log exponent for 6? _____

3. How are log of 3 and 6 like 2 and 4? _____

4. Find the exponent. What's the place value?   **log 30 = ?**

Place Value is ____   **Finish it.**

log 30 = ____

5. Find the exponent. What's the place value?   **log 6000 = ?**

Place Value is ____   **Finish it.**

log 6000 = ____

#3 Find the log values.   Calculator?
                                   yes   no

1.   log 2,000 = ____   log 300 = ____

2.   log 60,000 = ____   log 4,000,000 = ____

3.   log 4,000 = ____   log 30,000 = ____

4.   log 600,000 = ____   log 20 = ____

5.   log 2,000,000 = ____   log 4 = ____

6.   log 2 = ____   log 20 = ____

7.   log 60 = ____   log 300,000 = ____

8.   log 4,000,000 = ____   log 600 = ____

Review

1. Name the log of 2. How can you remember it? _____   Calculator?
                                                                         yes   no
2. How can you remember the log of 4? _____

3. Name 2 parts to a log exponent. **Log 20 = 1.30** _____
   _____

4. What is the log exponent for 3? _____

5. What is the log exponent for 6? _____

6. How are log of 3 and 6 like 2 and 4? _____

7.   log 3,000,000 = ____   log 40,000 = ____

8.   log 200,000 = ____   log 6,000,000 = ____

9.   log 400 = ____   log 20,000 = ____

10. log 300,000 = ____   log 60 = ____

Ch 1 Ls 5 Log Decimals for 7, 8, 9, and 5   285

_____ #1 #2 ____/10 #3 #4 ____/15 R ____/14 Total ____/39 _____
            Name   Checker

#1   1. What's the exponent for 7? _____
         2. How does 8 remind you of it's log? _____
         3. How is 9 different from the other log exponents? _____
         4. How do you use log 7 to find log 8 and 9? _____
         5. What's the exponent for 5? _____

   6. What's the log exponent for 700?   **log 700 = ?**

                                              log 700 = ____

   7. What's the log exponent for 8000?   **log 8000 = ?**

                                              log 8000 = ____

   8. What's the log exponent for 90?   **log 90 = ?**

                                              log 90 = ____

   9. What's the log exponent for 500?   **log 500 = ?**

                                              log 500 = ____

                    #2   Find the exponents.
   1. **log 8000**   **log 700,000**   **log 900**

   log _____ = ____   log _____ = ____   log _____ = ____

   2. **log 50**   **log 60,000**   **log 3,000**

   log _____ = ____   log _____ = ____   log _____ = ____

#3 Find the log values.   Calculator?
                            yes   no

1. log 50,000 = ____   log 8,000,000 = ____

2. log 9,000 = ____   log 70,000 = ____

3. log 800,000 = ____   log 90 = ____

4. log 5,000,000 = ____   log 700 = ____

5. log 500 = ____   log 80 = ____

6. log 70 = ____   log 900,000 = ____

7. log 80,000 = ____   log 7,000,000 = ____

8. log 9,000 = ____   log 500 = ____

Review 1. What's the exponent for 7? _____   Calculator?
        2. How does 8 remind you of it's log? _____   yes   no
        3. How is 9 different from the other log exponents? _____
        4. How do you use log 7 to find log 8 and 9? _____
        5. What's the exponent for 5? _____

6. log 6,000 = ____   log 50,000 = ____

7. log 90,000 = ____   log 4,000,000 = ____

8. log 7,000 = ____   log 300,000 = ____

9. log 200,000 = ____   log 800 = ____

10. log 40,000 = ____   log 600,000 = ____

Ch 1 Ls 6 Find a log real number from a log decimal.   287

_____ #1 #2 ____/13 #3 #4 ____/16 R ___/14 Total ____/ 43 _____
　　　　　Name　　　　　　　　　　　　　　　　　　　　　　　　　　　　　　　　　　　　Checker

#1 1. What does the decimal of the exponent, 2.48 show? _____

2. What does 2 from the exponent, 2.48 show? _____

3. What is the real number for the exponent 2.48? _____

4. What's the decimal part for exponent 3.90?   **log ? = 3.90**

　　　　　　　　　　　　　　　　　0.90 is exponent for _____   What's the real number?

　　　　　　　　　　　　　　log _____ = 3.90

5. What's the decimal part for exponent 2.48?   **log ? = 2.48**

　　　　　　　　　　　　　　　　　2.48 is exponent for _____   What's the real number?

　　　　　　　　　　　　　　log _____ = 2.48

6. What's the real number for 1.70?　　**log ? = 1.70**

　　　　　　　　　　　　　　log _____ = 1.70

7. What's the real number for 4.95?　　**log ? = 4.95**

　　　　　　　　　　　　　　log _____ = 4.95

#2 Find these real numbers.

1. **log ? = 1.30**　　　　　**log ? = 2.48**　　　　　**log ? = 1.95**

　log _____ = 1.30　　　log _____ = 2.48　　　log _____ = 1.95

2. **log ? = 2.70**　　　　　**log ? = 1.90**　　　　　**log ? = 3.85**

　log _____ = 2.70　　　log _____ = 1.90　　　log _____ = 3.85

#3 Find the value of each logarithm.   Calculator? yes no

1. log _____ = 1.30   log _____ = 2.60
2. log _____ = 3.48   log _____ = 5.85
3. log _____ = 2.90   log _____ = 4.70
4. log _____ = 1.95   log _____ = 2.30
5. log _____ = 3.78   log _____ = 2.95
6. log _____ = 4.30   log _____ = 5.85
7. log _____ = 2.78   log _____ = 2.90
8. log _____ = 3.70   log _____ = 1.78

Review 1. What does the decimal of the exponent, 2.48 show? _____   Calculator? yes no
2. What does 2 from the exponent, 2.48 show? _____
3. What is the real number for the exponent 2.48? _____

4. log _____ = 5.48   log _____ = 2.85
5. log _____ = 4.90   log _____ = 3.70
6. log _____ = 5.30   log _____ = 1.60
7. log _____ = 3.95   log _____ = 4.78
8. log _____ = 2.48   log _____ = 3.90
9. log _____ = 3.60   log _____ = 4.95
10. log _____ = 5.78   log _____ = 6.85

Review Problems 289

_____ #1 #2 #3 #4 ____/ 42 #5 #6 #7 ____/ 38 Total ____/ 80
      Name

1. Logarithm _____
2. Base _____
3. Exponent _____
4. Real Number _____

#2 Solve these exponents.

Calculator? yes no

1. $2^{\_} = 16$ $7^{\_} = 49$ $2^{\_} = 8$ $3^{\_} = 9$
2. $3^{\_} = 27$ $6^{\_} = 36$ $4^{\_} = 16$ $5^{\_} = 25$
3. $4^{\_} = 64$ $2^{\_} = 32$ $6^{\_} = 216$ $2^{\_} = 64$
4. $3^{\_} = 81$ $2^{\_} = 128$ $8^{\_} = 648$ $10^{\_} = 1000$
5. $2^{\_} = 256$ $5^{\_} = 625$ $4^{\_} = 256$ $9^{\_} = 81$

#3 Find the real number.

Calculator? yes no

1. log _____ = 3   log _____ = 8
2. log _____ = 2   log _____ = 5
3. log _____ = 7   log _____ = 1
4. log _____ = 4   log _____ = 6
5. log _____ = 10   log _____ = 9

#4 Change these exponents to log form.

Calculator? yes no

1. 10^5 is log _____ = ____ 10^6 is log _____ = ____
2. 10^7 is log _____ = ____ 10^9 is log _____ = ____
3. 10^3 is log _____ = ____ 10^8 is log _____ = ____
4. 10^2 is log _____ = ____ 10^4 is log _____ = ____

#5 Find the log values.   Calculator? yes no

1. log 200,000 = _____   log 6,000,000 = _____

2. log 3,000,000 = _____   log 40,000 = _____

3. log 400 = _____   log 20,000 = _____

4. log 6,000 = _____   log 50,000 = _____

5. log 300,000 = _____   log 60 = _____

6. log 7,000 = _____   log 300,000 = _____

7. log 90,000 = _____   log 4,000,000 = _____

8. log 40,000 = _____   log 600,000 = _____

9. log 200,000 = _____   log 800 = _____

#6 Find the value of each logarithm.   Calculator? yes no

1. log _____ = 2.90   log _____ = 4.70

2. log _____ = 4.48   log _____ = 3.85

3. log _____ = 5.95   log _____ = 4.78

4. log _____ = 4.30   log _____ = 2.60

5. log _____ = 1.60   log _____ = 3.95

6. log _____ = 3.48   log _____ = 4.90

#7 Solve these questions.

1. Name the log of 2 and 4? _____   Calculator? yes no

2. What is the log exponent for 3 and 6? _____

3. How do you use log 7 to find log 8 and 9? _____

4. What's the exponent for 5? _____

Ch 2 Ls 1 Add Logarithms   291

_____ #1 #2 ____ / 11 #3 ____ / 9 R ____ /14 Total ____ / 34 _____
         Name   Checker

#1 1. Log plus log equals what kind of answer? **log 4 + log 2** _____
2. How do you find the answer when you add logs? _____
3. What's the 2nd kind of answer adding logs makes? _____
4. How do you use log exponents to add logs? _____
5. 3 things to remember for add logs law. 1. Same _____
   2. Log + log = a _____ 3. _____ Rule backwards

#2 1. How do you find a log answer?   **log 2 + log 3**

_____. Find the exponents.   = log ____

What's the log with the exponent?   ____ + ____ = **?**

log ____ = ____

2. What's the log answer?   **log 4 + log 5**

Find the exponents.   = log ____

What's the log with the exponent?   ____ + ____ = **?**

log ____ = ____

3. What's the log answer?   **log 3 + log 6**

Find the exponents.   = log ____

What's the log with the exponent?   ____ + ____ = **?**

log ____ = ____

#3 Find the log answer, then the exponent answer.   Calculator?
                                                         yes   no

1. log 2 + log 3   log 5 + log 6   log 3 + log 4
    = log ___   = log ___   = log ___
    ___ + ___ = ___   ___ + ___ = ___   ___ + ___ = ___

2. log 5 + log 9   log 4 + log 7   log 2 + log 6
    = log ___   = log ___   = log ___
    ___ + ___ = ___   ___ + ___ = ___   ___ + ___ = ___

3. log 3 + log 3   log 4 + log 8   log 7 + log 8
    = log ___   = log ___   = log ___
    ___ + ___ = ___   ___ + ___ = ___   ___ + ___ = ___

Review 1. Log plus log equals what kind of answer? log 4 + log 2 _____   Calculator?
        2. How do you find the answer when you add logs? _____   yes   no
        3. What's the 2nd kind of answer adding logs makes? _____
        4. How do you use log exponents to add logs? _____
        5. 3 things to remember for add logs law. 1. Same _____
           2. Log + log = a _____ 3. _____ Rule backwards

6. log 4 + log 6   log 3 + log 9   log 8 + log 9
    = log ___   = log ___   = log ___
    ___ + ___ = ___   ___ + ___ = ___   ___ + ___ = ___

7. log 2 + log 8   log 4 + log 5   log 6 + log 6
    = log ___   = log ___   = log ___
    ___ + ___ = ___   ___ + ___ = ___   ___ + ___ = ___

Ch 2 Ls 2 Add Larger Numbers with Logarithms 293

_____ #1 #2 ____/ 7 #3 ____/ 6 R ____/ 6 Total ____/ 19 _____
        Name   Checker

#1 1. What's the 1st step to find the log answer for larger numbers? _____
_____

2. What's after the 1st digits? _____

3. What's the 1st step to find an exponent answer? _____

4. What happens to the exponents? _____

#2 1. How do you find a log answer?   log 2,000 + log 3,000

_____. Find the exponents.   = log _____

What's the log with the exponent?   ____ + ____ = ?

log _____ = ____

2. Find a log answer.   log 5,000 + log 40,000

Find the exponents.   = log _____

What's the log with the exponent?   ____ + ____ = ?

log _____ = ____

3. Solve for a log answer.   log 40,000
                             + log 800,000

= _____   What are the
                           exponents?

Find an exponent answer.   _____ + _____

log _____ = _____

#3 Find the log answer, then a log exponent.   Calculator? yes   no

1. log 2,000   log 3,000
    + log 5,000   + log 6,000
    ─────────   ─────────

   log _____   log _____

   ____ + ____ = ____   ____ + ____ = ____

2. log 3,000   log 70,000
    + log 50,000   + log 80,000
    ─────────   ─────────

   log _____   log _____

   ____ + ____ = ____   ____ + ____ = ____

3. log 9,000   log 400,000
    + log 300,000   + log 3,000,000
    ─────────   ─────────

   log _____   log _____

   ____ + ____ = ____   ____ + ____ = ____

Review 1. What's the 1st step to find the log answer for larger numbers? _____   Calculator?
        _____   yes   no

        2. What's after the 1st digits? _____

        3. What's the 1st step to find an exponent answer? _____

        4. What happens to the exponents? _____

        5. log 30,000   log 500,000
            + log 70,000   + log 2,000,000
            ─────────   ─────────

           log _____   log _____

           ____ + ____ = ____   ____ + ____ = ____

Ch 2 Ls 3 Subtract Logarithms   295

_____ #1 #2 ____/ 9 #3 ____/ 9 R ___/ 10 Total ____/ 28 _____
   Name   Checker

#1 1. Log - Log equals what kind of answer? _____
     2. What do you use to subtract Log - Log? _____
     3. How does it find a log answer for **Log 8 - log 4**? _____
     4. How do you find an exponent answer? _____
     5. How do you solve it? _____
     6. What are 2 kinds of answers for subtract logs? _____

#2 1. How do you find a log answer?   **log 6 - log 3**

           _____. Find the exponents.   = log ____

           What's the exponent answer?   ____ - ____ = ?

                                                log ____ = ____

     2. What's the log answer?   **log 15 - log 3**

           Find the exponents.   = log ____

           What's the exponent answer?   ____ - ____ = ?

                                                log ____ = ____

     3. What's the log answer?   **log 20 - log 2**

           Find the exponents.   = log ____

           What's the exponent answer?   ____ - ____ = ?

                                                log ____ = ____

#3 Find the log answer, then the exponent answer.   Calculator?
yes   no

1. log 20 - log 4   log 60 - log 20   log 50 - log 10
    = log ___   = log ___   = log ___
    ___ - ___ = ___   ___ - ___ = ___   ___ - ___ = ___

2. log 9 - log 3   log 30 - log 6   log 35 - log 5
    = log ___   = log ___   = log ___
    ___ - ___ = ___   ___ - ___ = ___   ___ - ___ = ___

3. log 25 - log 5   log 40 - log 5   log 15 - log 3
    = log ___   = log ___   = log ___
    ___ - ___ = ___   ___ - ___ = ___   ___ - ___ = ___

Review 1. Name 2 things you need to subtract logs. _____   Calculator?
        2. What rule subtracts logs? _____   yes   no
        3. How does it find a log answer for **Log 8 - log 4**? _____
        4. What are 2 kinds of answers for subtract logs? _____

5. log 30 - log 6   log 35 - log 7   log 45 - log 9
    = log ___   = log ___   = log ___
    ___ - ___ = ___   ___ - ___ = ___   ___ - ___ = ___

6. log 21 - log 3   log 25 - log 5   log 20 - log 2
    = log ___   = log ___   = log ___
    ___ - ___ = ___   ___ - ___ = ___   ___ - ___ = ___

Ch 2 Ls 4 Find the log exponent for fractions over 1.   297

_____ #1 #2 ____ / 7   #3 ____ / 9   R ____ / 7   Total ____ / 23 _____
Name   Checker

#1 1. How do you solve the real number of a log fraction over 1? _____
_____

2. How do you use fraction exponents for an exponent answer? _____
3. What's the 1st step for log of a mixed number? _____
4. How do you find the log exponent for a decimal? _____

#2 1. Find both log exponents for 20 and 5.   log $\frac{20}{5}$

Solve the log exponent.   log [_____]

Solve the real number.
Find the log answers.   ____ - ____ = ____

log ___ = ___

2. Find the log exponent.   log $1\frac{3}{5}$ = ?
What's the 1st step?

Solve the log exponent.   log _____

Solve the real number.
Find the log answers.   ____ - ____ = ____

log ___ = ___

3. Find the log exponent.   log $2\frac{3}{4}$ = ?
What's the 1st step?

Solve the log exponent.   log _____

Solve the real number.
Find the log answers.   ____ - ____ = ____

log ___ = ___

#3 Find the log exponents, then find each logarithm.   Calculator? yes   no

1. $\log \dfrac{30}{5}$ $\log \dfrac{15}{3}$ $\log \dfrac{42}{6}$

   log [____]   log [____]   log [____]

   log ___ = ___   log ___ = ___   log ___ = ___

2. $\log 1\dfrac{2}{5} =$ ___ $\log 2\dfrac{1}{3} =$ ___ $\log 4\dfrac{1}{2} =$ ___

   log [____]   log [____]   log [____]

   log ___ = ___   log ___ = ___   log ___ = ___

3. $\log 2\dfrac{2}{3} =$ ___ $\log 5\dfrac{4}{5} =$ ___ $\log 2\dfrac{1}{4} =$ ___

   log [____]   log [____]   log [____]

   log ___ = ___   log ___ = ___   log ___ = ___

Review 1. How do you solve the real number of a log fraction over 1? _____   Calculator? yes   no
_____

2. How do you use fraction exponents for an exponent answer? _____
3. What's the 1st step for log of a mixed number? _____
4. How do you find the log exponent for a decimal? _____

5. $\log \dfrac{56}{8}$ $\log 3\dfrac{1}{5} =$ ___ $\log 4\dfrac{1}{3} =$ ___

   log [____]   log [____]   log [____]

   log ___ = ___   log ___ = ___   log ___ = ___

Ch 2 Ls 5 Log Exponents for Fractions Less Than 1   299

_____ #1 #2 ____/ 7   #3 ____/ 6   R ___/ 9   Total ____/ 22   _____
            Name   Checker

#1 1. What log has exponent 0? _____

2. What are the log exponents for log 1 4th? _____

3. What are the log exponents for log 3 4ths? _____

#2 1. What are log exponents for 1 and 5? $\log \dfrac{1}{5}$

  Subtract.
  What's the log form? $\log \boxed{}$ ___ - ___ = ___

  $\log$ ___ = ___

2. What are log exponents for 1 and 3? $\log \dfrac{1}{3}$

  Subtract.
  What's the log form? $\log \boxed{}$ ___ - ___ = ___

  $\log$ ___ = ___

3. What are log exponents for 2 and 5? $\log \dfrac{2}{5}$

  Subtract.
  What's the log form? $\log \boxed{}$ ___ - ___ = ___

  $\log$ ___ = ___

4. What are log exponents for 3? $\log 0.3$

  Subtract.
  What's the log form? $\log \boxed{}$ ___ - ___ = ___

  $\log$ ___ = ___

#3 Find the log exponents, then find each logarithm.   Calculator? yes no

1. $\log \dfrac{4}{5}$ $\log \dfrac{2}{3}$ $\log \dfrac{1}{2}$

___ - ___ = ___   ___ - ___ = ___   ___ - ___ = ___

log ___ = ___   log ___ = ___   log ___ = ___

2. log 0.4   log 0.05   log 0.007

log [_____]   log [_____]   log [_____]

___ - ___ = ___   ___ - ___ = ___   ___ - ___ = ___

log ___ = ___   log ___ = ___   log ___ = ___

Review 1. What log has exponent 0? _____   Calculator? yes no

2. What are the log exponents for log 1 4th? _____

3. What are the log exponents for log 3 4ths? _____

4. $\log \dfrac{1}{4}$ $\log \dfrac{2}{7}$ $\log \dfrac{1}{3}$

___ - ___ = ___   ___ - ___ = ___   ___ - ___ = ___

log ___ = ___   log ___ = ___   log ___ = ___

5. log 0.6   log 0.02   log 0.008

log [_____]   log [_____]   log [_____]

___ - ___ = ___   ___ - ___ = ___   ___ - ___ = ___

log ___ = ___   log ___ = ___   log ___ = ___

Ch 2 Ls 6 Find Log Exponents for Decimals 301

_____ #1 #2 ____/ 8 #3 ____/ 6 R ____/ 9 Total ____/ 23 _____
           Name   Checker

#1 1. **log 0.8** What's the 1st step for a log exponent of a decimal? _____
2. **log 0.8** What happens to a log exponent to subtract 10ths? _____
3. Fractions less than 1 make what kind of exponent answer? _____

#2 1. What fraction does this make?   **log 0.006**

Subtract.
What's the log form?   log [———]   ___ - ___ = ___

                                log ____ = ____

2. What are log exponents for 5?   **log 0.005**

Subtract.
What's the log form?   log [———]   ___ - ___ = ___

                                log ____ = ____

3. What fraction does this make?   **log x = - 1.30**

                        log [———]   ___ - ___ = ___

4. What fraction does this make?   **log x = - 2.70**

                        log [———]   ___ - ___ = ___

5. What fraction does this make?   **log x = - 3.48**

                        log [———]   ___ - ___ = ___

#3 Solve the log exponents for the real numbers.   Calculator? yes no

1. log x = -1.30   log x = -1.48   log x = -2.90

   log ____   log ____   log ____

   ___ - ___ = ___   ___ - ___ = ___   ___ - ___ = ___

   log ___ = ___   log ___ = ___   log ___ = ___

2. log x = -3.60   log x = -2.95   log x = -1.70

   log ____   log ____   log ____

   ___ - ___ = ___   ___ - ___ = ___   ___ - ___ = ___

   log ___ = ___   log ___ = ___   log ___ = ___

Review 1. **log 0.8** What's the 1st step for a log exponent of a decimal? _____ Calculator? yes no

2. **log 0.8** What does log exponent to subtract 10ths? _____

3. Fractions less than 1 make what kind of exponent answer? _____

4. log x = -2.30   log x = -1.78   log x = -3.85

   log ____   log ____   log ____

   ___ - ___ = ___   ___ - ___ = ___   ___ - ___ = ___

   log ___ = ___   log ___ = ___   log ___ = ___

5. log x = -1.95   log x = -1.75   log x = -2.70

   log ____   log ____   log ____

   ___ - ___ = ___   ___ - ___ = ___   ___ - ___ = ___

   log ___ = ___   log ___ = ___   log ___ = ___

Review Problems 303

_____ #1 #2 #3 ____ / 19   #4 #5 ____ / 18 Total ____ / 37
Name

1. Add Logarithms _____
2. Overlap _____
3. Subtract Logarithms _____
4. Fraction Logs less than 1 _____
5. Decimal Logaritms _____

#2 Find the log answer, then the exponent answer.   Calculator?
                                                             yes   no

1.   log 5 + log 4   log 2 + log 6   log 7 + log 5
     = log ___   = log ___   = log ___
     ___ + ___ = ___   ___ + ___ = ___   ___ + ___ = ___

2.   log 4 + log 9   log 5 + log 8   log 7 + log 4
     = log ___   = log ___   = log ___
     ___ + ___ = ___   ___ + ___ = ___   ___ + ___ = ___

3.   log 3,000   log 6,000
       + log 8,000   + log 8,000
         ─────────   ─────────
    log _____   log _____
        ___ + ___ = ___   ___ + ___ = ___

#3 Do the same with these subtraction logs.   Calculator?
                                                         yes   no

1.   log 30 - log 5   log 80 - log 20   log 60 - log 10
     = log ___   = log ___   = log ___
     ___ - ___ = ___   ___ - ___ = ___   ___ - ___ = ___

2.   log 8 - log 4   log 40 - log 5   log 80 - log 40
     = log ___   = log ___   = log ___
     ___ - ___ = ___   ___ - ___ = ___   ___ - ___ = ___

#4 Find the log exponents, then find each logarithm.   Calculator? yes no

1. $\log \dfrac{40}{5}$ $\log \dfrac{27}{3}$ $\log \dfrac{30}{6}$

   log ___ = ___   log ___ = ___   log ___ = ___

2. $\log \dfrac{3}{4}$ $\log \dfrac{6}{7}$ $\log \dfrac{4}{5}$

   ___ - ___ = ___   ___ - ___ = ___   ___ - ___ = ___

   log ___ = ___   log ___ = ___   log ___ = ___

3. $\log 1\dfrac{3}{5} =$ ___ $\log 1\dfrac{3}{4} =$ ___ $\log 2\dfrac{2}{3} =$ ___

   ___ - ___ = ___   ___ - ___ = ___   ___ - ___ = ___

   log ___ = ___   log ___ = ___   log ___ = ___

4. log 0.7   log 0.02   log 0.008

   ___ - ___ = ___   ___ - ___ = ___   ___ - ___ = ___

   log ___ = ___   log ___ = ___   log ___ = ___

#5 Solve the log exponents for the real numbers.

1. log x = -2.70   log x = -3.68   log x = -1.85   Calculator? yes no

   log ⬜   log ⬜   log ⬜

   ___ - ___ = ___   ___ - ___ = ___   ___ - ___ = ___

   log ___ = ___   log ___ = ___   log ___ = ___

2. log x = -2.30   log x = -3.95   log x = -1.30

   log ⬜   log ⬜   log ⬜

   ___ - ___ = ___   ___ - ___ = ___   ___ - ___ = ___

   log ___ = ___   log ___ = ___   log ___ = ___

Ch 3 Ls 1 Estimate log exponents for 2 digit numbers. 305

_____ #1 #2 ____/ 9 #3 ____/ 10 R ___/ 4 Total ____/ 23 _____
           Name   Checker

#1 1. What's the 1st step to estimate logs of 2 digit numbers? _____
      2. What's next after finding the 1st digit? _____
      3. How do you estimate an exponent? _____
      4. How does log exponent change for log 2500? _____

#2 1. What are the perfect logs around 3.5?   **log 35**

        Estimate the exponent. log 3 = ____ log 4 = ____

              log 35 = ____

    2. What are the perfect logs around 3.5?   **log 75**

        Estimate the exponent. log 7 = ____ log 8 = ____

              log 75 = ____

    3. What are the perfect logs around 3.5?   **log 15**

        Estimate the exponent. log 1 = ____ log 2 = ____

              log 15 = ____

    4. All 1 step. What's the exponent for 65?   **log 65**

              log 65 = ____

    5. All 1 step. What's the exponent for 95?   **log 95**

              log 95 = ____

#3 Find the perfect logs, then estimate the exponent.   Calculator?
                                                              yes   no

1. **log 3500**   **log 4500**

    log 3 = ___   log 4 = ___   log 4 = ___   log 5 = ___

    log 3500 = ___   **log 4500 =** ___

2. **log 150**   **log 850**

    log 1 = ___   log 2 = ___   log 8 = ___   log 9 = ___

    log 150 = ___   **log 850 =** ___

3. **log 4500**   **log 250**

    log 4 = ___   log 5 = ___   log 2 = ___   log 3 = ___

    log 4500 = ___   **log 250 =** ___

4. **log 7500**   **log 6500**

    log 7 = ___   log 8 = ___   log 6 = ___   log 7 = ___

    log 7500 = ___   **log 6500 =** ___

5. **log 250**   **log 350**

    log 2 = ___   log 3 = ___   log 3 = ___   log 4 = ___

    log 250 = ___   **log 350 =** ___

Review 1. What's the 1st step to estimate logs of 2 digit numbers? _____   Calculator?
                                                                                    yes   no
2. What's next after finding the 1st digit? _____

3. How do you estimate an exponent? _____

4. How does log exponent change for log 2500? _____

Ch 3 Ls 2 Use logs with scientific notation. 307

_____ #1 #2 ____/ 8 #3 ____/ 12 R ___/ 12 Total ____/ 32 _____
         Name   Checker

#1 1. What's the scientific notation for 800? _____

2. How does it make a logarithm? _____

3. What does the 2nd digit do to the log exponent? _____

#2 1. What's the number for this scientific notation? 6×10^3

Estimate it. What is the log exponent? _____

_____

2. What's the 1st step to find this number? 4×10^8

Write the _____. What's left? _____

_____

3. What's the number for this scientific notation? 6.5×10^3

Estimate it. What is the log exponent? _____

_____

4. What's the 1st step to find this number? 4.5×10^4

Write the _____. What's left? _____

_____

5. What's the 1st step to find this number? 5.5×10^5

Write the _____. What's left? _____

_____

#3 Decide the log of each scientific notation.   Calculator? yes no

1. 5×10^2 8×10^4 9×10^3

    log ____ = ____   log ____ = ____   log ____ = ____

2. 7×10^3 6×10^3 2×10^2

    log ____ = ____   log ____ = ____   log ____ = ____

3. 3.5×10^4 4.5×10^2 1.5×10^3

    log ____ = ____   log ____ = ____   log ____ = ____

4. 4.5×10^2 5.5×10^4 8.5×10^3

    log ____ = ____   log ____ = ____   log ____ = ____

Review 1. What's the scientific notation for 800? _____   Calculator? yes no
2. How does it make a logarithm? _____
3. What does the 2nd digit do to the log exponent? _____

4. 3×10^2 6×10^4 4×10^3

    log ____ = ____   log ____ = ____   log ____ = ____

5. 1.5×10^3 5.5×10^2 6.5×10^4

    log ____ = ____   log ____ = ____   log ____ = ____

6. 7.5×10^3 4.5×10^2 3.5×10^4

    log ____ = ____   log ____ = ____   log ____ = ____

Ch 3 Ls 3 Use Power Law with Logarithms   309

_____ #1 #2 ____ / 7 #3 ____ / 9 R ___ / 7 Total ____ / 23 _____
        Name   Checker

#1 1. How do you solve the real number with an exponent? _____

2. How do you use a Power Log with an exponent? _____

3. Name 2 ways to solve a log with an exponent. _____

_____

4. How do you solve an exponent to an exponent? _____

    #2 1. Find the power log. $\log 6^2$

           Finish the log exponent.   ___ x log ___

                                        log ___ = ___

      2. Find the power log. $\log (6^2)^2$

           What's the exponent for 6?   ___ x log ___

           Finish the log exponent.   ___ x log ___

                                        log ___ = ___

      3. Find the power log. $\log (3^2)^2$

           What's the exponent for 6?   ___ x log ___

           Finish the log exponent.   ___ x log ___

                                        log ___ = ___

#3 Use the power rule, then find the log exponent.   Calculator? yes   no

1. $\log 7^2$ $\log 9^2$ $\log 2^2$

___ x log ___   ___ x log ___   ___ x log ___

log ___ = ___   log ___ = ___   log ___ = ___

2. $(\log 4^2)^2$ $(\log 3^2)^2$ $(\log 8^2)^2$

___ x log ___   ___ x log ___   ___ x log ___

log ___ = ___   log ___ = ___   log ___ = ___

3. $(\log 5^2)^3$ $(\log 8^4)^2$ $(\log 6^3)^2$

___ x log ___   ___ x log ___   ___ x log ___

log ___ = ___   log ___ = ___   log ___ = ___

Example $\log 6^2$

Review 1. How do you solve the real number with an exponent? _____   Calculator? yes no

2. How do you use a Power Log with an exponent? _____

3. Name 2 ways to solve a log with an exponent. _____

_____

4. How do you solve an exponent to an exponent? _____

5. $\log 4^2$ $(\log 3^2)^2$ $(\log 7^3)^2$

___ x log ___   ___ x log ___   ___ x log ___

log ___ = ___   log ___ = ___   log ___ = ___

Ch 3 Ls 4 Log Exponents for Square Roots 311

_____ #1 #2 ____/ 8 #3 ____/ 9 R ___/ 10 Total ____/ 27 _____
                Name   Checker

#1 1. What is the exponent form for any square root? _____
  2. How does it use a power log? _____
  3. What's the shortcut for log of a square root? _____
  4. Name 2 ways to solve a mixed exponent. _____
  _____

#2 1. Find the exponent. $\log \sqrt{3}$

    Find the power log. $\log$ ____

    Finish the log exponent.   ____ x log ____

                                             $\log$ ____ = ____

  2. Find the exponent. $\log 2^{\frac{2}{3}}$

    Find the power log. $\log$ ____

    Finish the log exponent.   ____ x log ____

                                             $\log$ ____ = ____

  3. All 1 step? What's the
     exponent for square root of 4? $\log \sqrt{4}$

                                             $\log \sqrt{4}$ = _____

  4. All 1 step? What's the
     exponent for square root of 4? $\log \sqrt{5}$

                                             $\log \sqrt{5}$ = _____

312

#3 Decide the log of each logarithm.

1. $\log \sqrt{3}$ $\log \sqrt{6}$ $\log \sqrt{8}$ Calculator? yes   no

   __ x log __   __ x log __   __ x log __

   log __ = __   log __ = __   log __ = __

2. $\log \sqrt{20}$ $\log \sqrt{40}$ $\log \sqrt{70}$

   __ x log __   __ x log __   __ x log __

   log __ = __   log __ = __   log __ = __

3. $\log 2^{\frac{2}{3}}$ $\log 5^{\frac{3}{4}}$ $\log 9^{\frac{2}{3}}$

   __ x log __   __ x log __   __ x log __

   log __ = __   log __ = __   log __ = __

Review
1. What is the exponent form for any square root? _____   Calculator? yes   no
2. How does it use a power log? _____
3. What's the shortcut for log of a square root? _____
4. Name 2 ways to solve a mixed exponent. _____

_____

5. $\log \sqrt{12}$ $\log \sqrt{50}$ $\log \sqrt{80}$

   __ x log __   __ x log __   __ x log __

   log __ = __   log __ = __   log __ = __

6. $\log 4^{\frac{2}{5}}$ $\log 3^{\frac{3}{4}}$ $\log 8^{\frac{2}{3}}$

   __ x log __   __ x log __   __ x log __

   log __ = __   log __ = __   log __ = __

Review Problems 313

_____ #1 #2 #3 ____ / 18 #4 #5 ____ / 18 Total ____ / 36
         Name

1. 2 Digit Logarithms _____
2. Sciemtific Notation Logarithms _____
3. Poweer Logarithm _____
4. Exponent to an Exponent _____
5. Radical Logarithm _____

#2 Find the perfect logs, then estimate the exponent.   Calculator?
                                                                 yes   no

1.   **log 250**   **log 750**

   log 2 = ___   log 3 = ___   log 7 = ___   log 8 = ___

        log 250 = _____   log 750 = _____

2.   **log 5500**   **log 8500**

   log 5 = ___   log 6 = ___   log 8 = ___   log 9 = ___

        log 5500 = _____   log 8500 = _____

#3 Decide the log of each scientific notation.   Calculator?
                                                              yes   no

1. 7×10^2 6×10^4 8×10^3

   log _____ = ___   log _____ = ___   log _____ = ___

2. 6.5×10^3 8.5×10^2 9.5×10^4

   log _____ = ___   log _____ = ___   log _____ = ___

3. 7.5×10^2 5.5×10^4 2.5×10^3

   log _____ = ___   log _____ = ___   log _____ = ___

#4 Use the power rule, then find the log exponent.   Calculator? yes   no

1. $\log 8^3$ $\log 7^4$ $\log 5^2$

___ x log ___   ___ x log ___   ___ x log ___

log ___ = ___   log ___ = ___   log ___ = ___

2. $(\log 2^2)^3$ $(\log 5^4)^2$ $(\log 7^2)^3$

___ x log ___   ___ x log ___   ___ x log ___

log ___ = ___   log ___ = ___   log ___ = ___

#5

1. $\log \sqrt{2}$ $\log \sqrt{5}$ $\log \sqrt{7}$ Calculator? yes no

___ x log ___   ___ x log ___   ___ x log ___

log ___ = ___   log ___ = ___   log ___ = ___

Decide the log of each logarithm.

2. $\log \sqrt{30}$ $\log \sqrt{60}$ $\log \sqrt{80}$

___ x log ___   ___ x log ___   ___ x log ___

log ___ = ___   log ___ = ___   log ___ = ___

3. $\log 8^{\frac{3}{5}}$ $\log 7^{\frac{3}{4}}$ $\log 5^{\frac{2}{3}}$

___ x log ___   ___ x log ___   ___ x log ___

log ___ = ___   log ___ = ___   log ___ = ___

4. $\log 6^{\frac{2}{3}}$ $\log 2^{\frac{5}{6}}$ $\log 4^{\frac{3}{5}}$

___ x log ___   ___ x log ___   ___ x log ___

log ___ = ___   log ___ = ___   log ___ = ___

Ch 4 Ls 1 Logarithms with Different Bases   315

_____ #1 #2 ____/ 7 #3 ____/ 12 R ___/ 11 Total ____/ 30 _____
            Name   Checker

#1 1. How does base 5 change logs from base 10? _____

2. $\log_5 x = 2$ What do different bases change? _____

#2 1. Change to logarithm form. $5^2 = x$

Solve for the log exponent. $\log\underline{}\underline{} = \underline{}$

$\log\underline{}\underline{} = \underline{}$

2. Change to logarithm form. $6^2 = x$

Solve for the log exponent. $\log\underline{}\underline{} = \underline{}$

$\log\underline{}\underline{} = \underline{}$

3. Change to logarithm form. $9^2 = x$

Solve for the log exponent. $\log\underline{}\underline{} = \underline{}$

$\log\underline{}\underline{} = \underline{}$

4. Solve for the log exponent. $20^2 = x$

$\log\underline{}\underline{} = \underline{}$

5. Solve for the log exponent. $25^2 = x$

$\log\underline{}\underline{} = \underline{}$

#3 Decide the log for each log.   Calculator? yes no

1. $5^3 = x$ $4^3 = x$ $3^3 = x$

   log __ = __   log __ = __   log __ = __

2. $8^2 = x$ $6^2 = x$ $9^2 = x$

   log __ = __   log __ = __   log __ = __

3. $17^2 = x$ $30^3 = x$ $15^2 = x$

   log __ = __   log __ = __   log __ = __

4. $12^3 = x$ $23^2 = x$ $40^3 = x$

   log __ = __   log __ = __   log __ = __

Review 1. How does base 5 change logs from base 10? _____   Calculator?
2. $\log_5 x = 2$ What do different bases change? _____   yes no

3. $20^3 = x$ $35^2 = x$ $18^2 = x$

   log __ = __   log __ = __   log __ = __

4. $75^2 = x$ $25^3 = x$ $50^3 = x$

   log __ = __   log __ = __   log __ = __

5. $100^2 = x$ $200^2 = x$ $300^2 = x$

   log __ = __   log __ = __   log __ = __

Ch 4 Ls 2 Change Other Bases with Base 10   317

_____ #1 #2 ____/ 6 #3 ____/ 6 R ___/ 5 Total _____/ 17 _____
Name   Checker

#1 1. How do you change base 5 to a base 10 log? _____
2. How is changing to another base besides 10 different? _____
3. How do you solve a fraction with log exponents? _____

#2 1. Change to base 10 log. $\log_5 25$

Find the exponents. $\dfrac{\log \_\_\_}{\log \_\_\_}$

Solve the log exponent. $\dfrac{\log \_\_\_}{\log \_\_\_}$

$\log 25 = $ _____

2. Change to base 3 log.
What are the base 10 logs? $\log_5 25$

Find the base 3 logs. $\dfrac{\log \_\_\_}{\log \_\_\_}$

$\dfrac{\log \_\_\_}{\log \_\_\_}$

3. Change to base 2 log.
What are the base 10 logs? $\log_4 40$

Find the base 2 logs. $\dfrac{\log \_\_\_}{\log \_\_\_}$

$\dfrac{\log \_\_\_}{\log \_\_\_}$

#3 Change Other Bases with Base   Calculator? yes no

1. Change to base 10 log. $\log_6 36$ $\log_2 12$

   Change to base 6 logs.   log ____   log ____
   How does it change?   log   log

                            log ____   log ____
                            log   log

2. Change to base 10 log. $\log_5 30$ $\log_5 40$

   Change to base 2 logs.   log ____   log ____
   How does it change?   log   log

                            log ____   log ____
                            log   log

3. Change to base 10 log. $\log_2 24$ $\log_3 36$

   Change to base 12 logs. log ____   log ____
   How does it change?   log   log

                            log ____   log ____
                            log   log

Review 1. How do you change base 5 to a base 10 log? _____   Calculator?
2. How is changing to another base besides 10 different? _____   yes no
3. How do you solve a fraction with log exponents? _____

4. Change to base 10 log. $\log_3 15$ $\log_5 45$

   Change to base 3 logs.   log ____   log ____
   How does it change?   log   log

                            log ____   log ____
                            log   log

Ch 4 Ls 3 Exponential Equations 319

_____ #1 #2 ____ / 7 #3 ____ / 4 R ____ / 5 Total ____ / 16 _____
            Name   Checker

#1 1. What makes an exponential equation? _____

2. It's 2 to the X power. What happens when X is -1? _____

3. What point is the same for all exponential equations? _____

4. What do exponential equation graphs look like? _____

5. How does a negative base change the graph? _____

#2 1. Solve for x is -2, 0, and 2. $y = 3^x$

Make the graphs. (-2, ____) (0, ____) (2, ____)

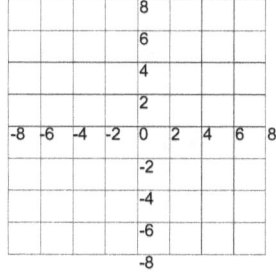

2. Solve for x is -2, 0, and 2. $y = -2^x$

Make the graphs. (-2, ____) (0, ____) (2, ____)

#3 Solve for x is -2, 0, and 2. Then estimate the exponents.   Calculator? yes no

1.

$y = 4^x$

-2, ___
1, ___

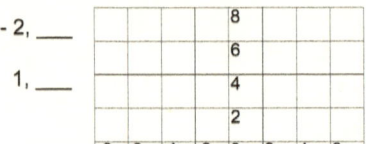

$y = -5^x$

-2, ___
1, ___

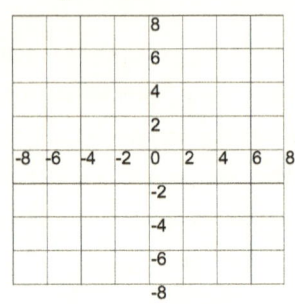

2.

$y = 10^x$

-2, ___
1, ___

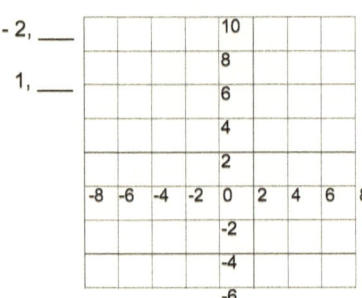

$y = -6^x$

-2, ___
1, ___

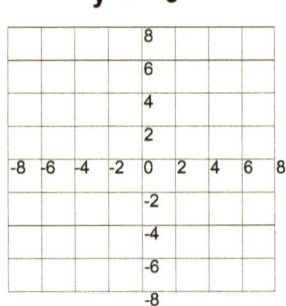

Review   1. What makes an exponential equation? _____   Calculator?
                                                                              yes   no
2. It's 2 to the X power. What happens when X is -1? _____

3. What point is the same for all exponential equations? _____

4. What do exponential equation graphs look like? _____

5. How does a negative base change the graph? _____

Ch 4 Ls 4 How Same Bases Solve Equations

_____ #1 #2 ___/ 6 #3 ___/ 6 R ___/ 3 Total ___/ 15 _____
Name   Checker

#1 1. How do same bases with variable exponents solve? _____

$$8 = 2^{n+1}$$

2. Which base changes if they are different? _____

$$2^n = \frac{1}{4}$$

3. What step changes a fraction to get same bases? _____

#2 1. What changes to get same bases? $9 = 3^{x+6}$

What equation solves it?   ___ $= 3^{x+6}$

Solve it for X.   _____

_____

2. What changes to get same bases? $2^n = \frac{1}{4}$

What equation solves it?   _____

Solve it for X.   _____

_____

3. What changes to get same bases? $4^n = \frac{1}{16}$

What equation solves it?   _____

Solve it for X.   _____

_____

#3 Find same bases, then mentally solve for x.   Calculator? yes no

1. $32 = 2^{x+2}$ $\qquad$ $64 = 4^{x+5}$

$\underline{\quad} = 2^{x+2}$ $\qquad$ $\underline{\quad} = 4^{x+5}$

_____   _____

_____   _____

2. $9 = 3^{x-1}$ $\qquad\qquad$ $27 = 3^{x+4}$

_____   _____

_____   _____

_____   _____

3. $125^n = \frac{1}{5}$ $\qquad\qquad$ $64^n = \frac{1}{4}$

_____   _____

_____   _____

_____   _____

Review 1. How do same bases with variable exponents solve? _____   Calculator? yes no

$$8 = 2^{n+1}$$

2. Which base changes if they are different? _____

$$2^n = \frac{1}{4}$$

3. What step changes a fraction to get same bases? _____

Review Problems **323**

_____ #1 ____/ 5   #2 ____/ 5   Total ____/ 10
               Name

                                                              Calculator?
                                                              yes   no

#1 1. Find a balance on a savings account balance if   _____
     the account starts with Rs 8000, has an annual   _____
     rate of 3 and the money is left in the account   _____
     for 8 years.   _____

Make an equation   _____
before solving it.

   2. In 1990, there were 800 cell phone subscribers   _____
      in the small town of Bieta. The number of
      subscribers increased by 60% per year after   _____
      2000. How many cell phone subscribers were
      in Biets in 2000?   _____

                                                             _____

                                                             _____

   3. Bacteria can double 2 cells for every one. If we _____
      start with 1 bacteria which can double every
      hour, how many bacteria will we have by the   _____
      end of 8 hours?
                                                             _____

                                                             _____

                                                             _____

   4. Each year a local country club sponsors a   _____
      bowling tournament. Play starts with 64
      participants. During each game, half of the   _____
      players are eliminated. How many players
      remain after 5 games?   _____

                                                             _____

                                                             1980
   5. The population of Zurbaga uses P = 23,000(1.04)^t _____
      where t is the number of years since 1980. What   What %?
      was the population in 1980? By what percent did   _____
      the population increase by each year? What is
      the end result in 1988?   _____

                                                             1988
                                                             _____

#2

1. You have inherited land that was purchased for Rs 20,000 in 1965. The land's value increased by approximately 5% per year. What is the approximate value of the land in the year 1995?

Calculator? yes   no

2. You deposit Rs 1000 in a bank account. Find the balance after 4 years if the account pays 2% annual interest compounded monthly and if it pays 4% annual interest compounded yearly.

3. An adult takes 325 mg of aspirin. Each hour, the amount of aspirin in the person's system decreases by about 30%. How much aspirin is left after 6 hours?

4. You buy a new computer for Rs 8000. The computer decreases by 40% annually. When will the computer have a value of Rs 2000?

5. You drink a beverage with 120 mg of caffeine. Each hour, the caffeine in your system decreases by about 12%. How long until you have 10mg of caffeine?

Review Problems   325

_____ #1 #2 #3 ____/ 17 #4 #5 ____/ 8 Total ____/ 25
           Name

1. Other Bases _____
2. Change to Base 10 _____
3. Change to Other Bases _____
4. Same Base Equations _____

#2   Decide the log for each log.   Calculator?
                                                                     yes   no

1. $5^3 = x$ $4^3 = x$ $3^3 = x$

     log ___ = ___   log ___ = ___   log ___ = ___
         ___   ___   ___

2. $8^2 = x$ $6^2 = x$ $9^2 = x$

     log ___ = ___   log ___ = ___   log ___ = ___
         ___   ___   ___

3. $17^2 = x$ $30^3 = x$ $15^2 = x$

     log ___ = ___   log ___ = ___   log ___ = ___
         ___   ___   ___

#3   Change Other Bases with Base   Calculator?
                                                                     yes   no
1. $\log_6 36$ $\log_2 12$

Change to base 10 log.   log _____   log _____
                          log   log

Change to base 4 logs.   log _____   log _____
How does it change?   log   log

2. $\log_5 30$ $\log_5 40$

Change to base 10 log.   log _____   log _____
                          log   log

Change to base 2 logs.   log _____   log _____
How does it change?   log   log

#4 1. $y = -7^x$ $y = 10^x$ Calculator? yes no

-2, ___   -2, ___
0, ___   0, ___
2, ___   2, ___

Solve for x is -2, 0, and 2.

Estimate the exponents.

2. $y = 8^x$ $y = -3^x$

-2, ___   -2, ___
0, ___   0, ___
2, ___   2, ___

#5 1. $8 = 2^{x-2}$ $49 = 7^{x+3}$ Calculator? yes no

Solve these equations.

2. $1000^n = \dfrac{1}{10}$ $27^n = \dfrac{1}{3}$

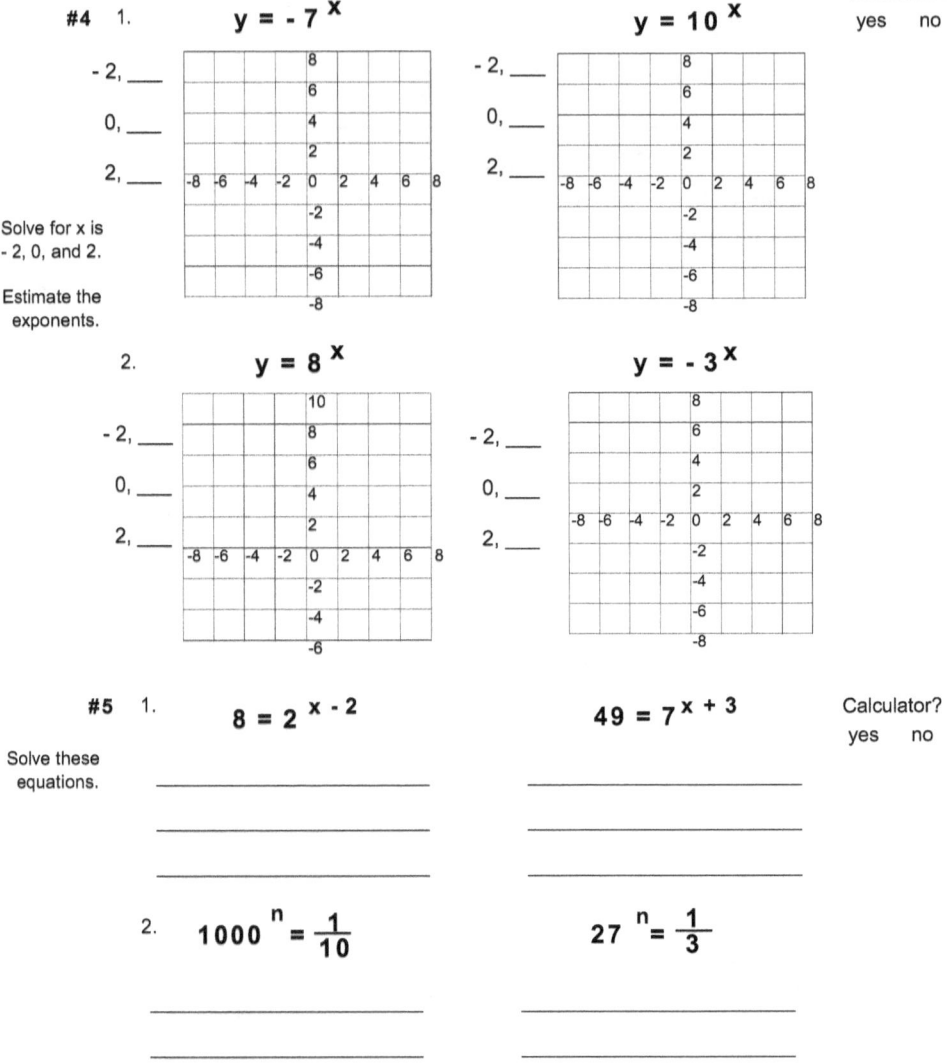

Ch 5 Ls 1, 2 Simple Logs, Log Equations with Antilogarithms   327

_____ #1 #2 ____ / 8   #3 ____ / 4 R ____ / 9 Total ____ / 21 _____
Name   Checker

#1 1. What 2 thngs do you need to solve a simple equation? _____

_____

2. Name the 1st step to solve the equation? _____

3. What's 1st to solve a simple equation? $\log_4 6 \, (\log_4 5) = \log_4 3x$

What solves the equation? _____

What's the answer? _____

_____

#2 1. What does an antilogarithm do? _____

2. When would you use an antilogarith? _____

3. What does antilog of a binomial make? $\log(x+1) = 2$ _____

4. What's the 1st step to an antilog? $\log_x 125 = 3$

What solves the equation? _____

What's the next step? _____

What's the answer? _____

5. What's the 1st step to an antilog? $\log(x-8) = 2$

What solves the equation? _____

What's the answer? _____

_____

#3 Solve these problems using antilogarithms.   Calculator? yes   no

1. $\log_5 3(\log_5 6) = \log_5 9x$ $2 \log 3 + 2 \log 6 = \log 3x$

_____   _____

_____   _____

_____   _____

2. $3 \log 3 (\log 4) = \log 3x$ $\log_7 3x + \log_7 4x = \log_7 24$

_____   _____

_____   _____

_____   _____

Review 1. What 2 thngs do you need to solve a simple equation? _____   Calculator? yes   no
_____

2. Name the 1st step to solve the equation? _____

3. What does an antilogarithm do? _____

4. When would you use an antilogarith? _____

5. What does antilog of a binomial make? log x + 1 = 2 _____

6. $\log (x - 570) = 6$ $\log (x - 300) = 5$

_____   _____

_____   _____

_____   _____

7. $\log_2 (x - 2) = 5$ $\log_3 (x - 12) = 2$

_____   _____

_____   _____

_____   _____

Ch 5 Ls 3 Take a Log of Both Sides   329

_____ #1 #2 ____/ 9 #3 ____/ 4 R ___/ 9 Total ____/ 22 _____
           Name   Checker

#1   1. What's the opposite of an antilogarithm? _____

2. What kind of problem uses log of both sides? _____

3. Why take the log of both sides? _____

4. What happens to the exponent? $2^{a+1} = 16$

What solves the equation? _____

What's the next step? _____

What's the answer? _____

_____

#2   1. What's the 1st question for an interest problem? _____

2. What's the growth formula? _____

3. What variable does it solve for? _____

4. Why does it take log of both sides? _____

5. What's the formula for it?   Your buying a bond for Rs 800. It pays 7% compounded annually until it reaches Rs 1000. How long until it matures?

What solves the equation? $V_n = P(1 + \text{rate})^N$

What's the next step?   _____

What's the answer?   _____

_____

_____

#3 1. $2^{a+2} = 32$ $6^{b+1} = 216$ Calculator? yes no

Solve these problems by taking the log of both sides.

2. $3^{x+1} = 243$ $4^{c+1} = 64$

Review
1. What's the opposite of an antilogarithm? _____   Calculator? yes no
2. What kind of problem uses log of both sides? _____
3. Why take the log of both sides? _____
4. What's the 1st question for an interest problem? _____
5. What's the growth formula? _____
6. What variable does it solve for? _____
7. Why does it take log of both sides? _____
8. A Rs 100,000 bond matures once a year. If it compounded annually at 8%, how long until it's reaches Rs 150,000.   If the problem starts at Rs 100,000, how long until it reaches Rs 200,000 at 10%?

Ch 5 Ls 4 Use Ma Rule with Log Equations   331

_____ #1 #2 ____ / 7   #3 ____ / 4   R ____ / 3   Total ____ / 14 _____
Name   Checker

#1   1. What rule solves 1 log multiplied? **log 5x = 1** _____

   2. How do you change 2 added logs to be 1 logarithm? _____

   3. How do you solve log of 2 binomials? **log(x - 1)(x + 2)** _____

   4. What solves the log of a trinomial? **log (x^2 + x - 2)** _____

#2   1. What's the 1st step to get 1 log?   **log 8x = 1**

          What solves the equation?   _____

          What's the next step?   _____

          What's the answer?   _____

                                            _____

   2. What's the 1st step to get 1 log?   **log 7 + log x = 1**

          What solves the equation?   _____

          What's the next step?   _____

          What's the answer?   _____

                                            _____

   3. Solve to find what the log's worth.
       What's the 1st step to get 1 log?   **log(x - 1) + log(x + 2) = 1**

          What solves the equation?   _____

          What's the next step?   _____

          What's the answer?   _____

                                            _____

#3 Solve these using the MA rule to find a log answer.   Calculator? yes no

1. **log 6x = 1**   **log 5x = 2**

Solve to find
the log's value. _____   _____

_____   _____

_____   _____

2. **log 3 + log x = 2**   **log 4 + log x = 1**

Solve to find
the log's value. _____   _____

_____   _____

_____   _____

3. **log(x + 1) + log(x + 4) = 1**   **log(2x + 3) + log(x + 1) = 1**

Solve to find
the log's value as
a real number. _____   _____

_____   _____

_____   _____

Review 1. How do you change 2 added logs to be 1 logarithm? _____   Calculator?
2. How do you solve log of 2 binomials? **log(x - 1)(x + 2)** _____   yes   no
3. What solves the log of a trinomial? **log (x^2 + x - 2)** _____

Ch 5 Ls 5 Use DS Rule with Log Equations   333

_____ #1 #2 ____ / 7 #3 ____ / 6 R ____ / 4 Total ____ / 17 _____
                Name   Checker

#1 1. How do you solve a log with a fraction? $\log \frac{x}{2}$ _____

   2. How do you solve $\log x - \log 2 = 1$? _____

   3. If 2 logs are subtracted, how can it be 1 logarithm? _____

   4. What's the 1st step? $\log \frac{5}{(x-2)} = 1$ _____

#2 1. What's the 1st step to get 1 log? $\log x - \log 8 = 1$

   How do you solve log of a fraction?   _____

   What's the next step?   _____

   What's the answer?   _____

                                          _____

   2. How do you change $\log 4/(x-1)$? $\log \frac{4}{(x-1)} = 1$

   What's the next step?   _____

   What's the answer?   _____

                                          _____

   3. How do you change 1/3 log 8? $2 \log_3 6 - 1/3 \log_3 8 = \log_3 x$

   What solves the equation?   _____

   What's the next step?   _____

   What's the answer?   _____

                                          _____

#3 Solve these problems by using the DS Rule. Calculator?
yes   no

1. log 8 - log(x - 2) = 1   log 4 - log(x - 3) = 1

   _____   _____
   _____   _____
   _____   _____
   _____   _____

2. log $\dfrac{2}{(x-2)}$ = 2   log $\dfrac{1}{(x-3)}$ = 1

   _____   _____
   _____   _____
   _____   _____
   _____   _____

3. log (x + 2) - log(x - 1) = 1   log (x + 4) - log(x - 3) = 1

   _____   _____
   _____   _____
   _____   _____
   _____   _____

Review 1. How do you solve a log with a fraction? log $\dfrac{x}{2}$ _____   Calculator?
yes   no
2. How do you solve **log x - log 2 = 1**? _____
3. If 2 logs are subtracted, how can it be 1 logarithm? _____
4. What's the 1st step? log $\dfrac{5}{(x-2)}$ = 1 _____

_____ #1 #2 ____/ 14 #3 #4 ____/ 8 Total ____/ 22
           Name

1. Same Log Equations _____
2. Antilogarithms _____
3. Both Sides Logarithms _____
4. DS Logarithms _____

#2 Solve these problems.   Calculator?
                                                         yes   no

1. $\log_3 3(\log_3 4) = \log_3 8x$ $2 \log 4 + 2 \log 7 = \log 5x$

   _____   _____
   _____   _____
   _____   _____

2. $\log(x - 240) = 3$ $\log(x - 420) = 4$

   _____   _____
   _____   _____
   _____   _____

3. $\log_2(x - 3) = 4$ $\log_3(x - 8) = 2$

   _____   _____
   _____   _____
   _____   _____

4. $3^{a + 3} = 81$ $5^{b + 1} = 625$

   _____   _____
   _____   _____
   _____   _____

#3 Solve these using the MA rule to find a log answer.   Calculator?
                                                                yes   no

1. $\log 5 + \log x = 3$ $\log 8 + \log x = 2$

Solve to find
the log's value.

_____   _____

_____   _____

_____   _____

2. $\log(x + 2) + \log(x + 6) = 4$ $\log(3x + 1) + \log(x + 2) = 2$

Solve to find
the log's value as
a real number.

_____   _____

_____   _____

_____   _____

#4 Solve these problems by using the DS Rule.   Calculator?
                                                      yes   no

1. $\log 8 - \log(x - 2) = 1$ $\log 4 - \log(x - 3) = 1$

_____   _____

_____   _____

_____   _____

_____   _____

2. $\log \dfrac{3}{(x - 4)} = 1$ $\log \dfrac{1}{(x - 6)} = 2$

_____   _____

_____   _____

_____   _____

_____   _____

Ch 6 Ls 1 Natural Numbers 337

_____ #1 #2 ____/ 8 #3 #4 ____/ 7 R ___/ 5 Total ____/ 20 _____
                Name   Checker

#1 1. What is the symbol and number for the natural number? _____

2. What's the shortcut to write **Log $_e$** ? _____

3. What 1 word tells about natural number? _____

4. How does it find continuous growth interest? _____

5. What do N and Y stand for? _____

#2 1. Solve these exponents. $e^3 \, (e^2)$

_____

2. Solve these exponents. $\dfrac{e^4}{e^2}$

_____

3. What is the formula for this?   A Rs 4000 account pays 8% per year compounded continuenty. What is it worth after 2 years?

Where does $100, 8%, and 2 go? $t = ne^{k(t)}$

What's the first step?   y = 4000e ——— ———

What's the next step?   y = 4000e ———

What are the next steps?   y = 4000(2.71) ———

What is the final answer? y = 4000(   )

                           y =

#3 Solve these exponents. Calculator? yes no

1. $e^3(e^2)$ $e^3(e^4)$ $e^5(e^2)$

_____ _____ _____

2. $\dfrac{e^3}{e^2}$ $\dfrac{e^2}{e^5}$ $\dfrac{e^3}{e^4}$

_____ _____ _____

#4 1. What is the formula for this?   A Rs 2000 account pays 9% per year continuenty compounded. What is it worth after 3 years? Calculator? yes no

Where does $200, 8%, and 3 go?   t = _____ e ― ―

What's the first step?   y = 2000e ― ―

What's the next step?   y = 2000e ―

What are the next steps?   y = 2000(2.71) ―

What is the final answer?   y = 2000(   )

           y =

Review 1. What is the symbol and number for the natural number? _____ Calculator? yes no

2. What's the shortcut to write **Log e** ? _____

3. What 1 word tells about natural number? _____

4. How does it find continuous growth interest? _____

5. What do N and Y stand for? _____

Ch 6 Ls 2 Growth Formua vs Compound Interest 339

_____ #1 #2 ____/ 9 #3 ____/ 3 R ___/ 5 Total ____/ 17 _____
   Name   Checker

#1 1. What is the growth formula? _____

2. What's the 1st step to solve it? $1000 = Pe^{10\%(4)}$ _____

3. What's the next step? _____

4. How does it finish? _____

5. What changes for a decay problem? $t = ne^{k(t)}$ _____

#2 1. What formula does it use?   You need Rs 6000 in 4 years. If an account pays 10% compounded continuecy per year. How much does it start with to get that?

Which numbers go where?   _____

What's the 1st step to solve it?   _____

                                 _____

Estimate the answer.   _____

                                 _____

2. What formula does this use?   A city of 4000 people is losing 5% per year. After how 4 years, how many are there left?

Which numbers go where?   _____

What's the 1st step to solve it?   _____

                                 _____

Estimate the answer.   _____

#3 Solve these story problems.   Calculator?
                                                                    yes   no

1. **Buy a bond for Rs 4000. At 5% compounded annually, how much is it worth after 2 years?**

   _____

   _____

   _____

   _____

   _____

2. **ABC Company bought a computer company for Rs 700,000. If declines at a rate of 10% per year, how much is it worth after 3 years?**

   _____

   _____

   _____

   _____

   _____

3. **A Rs 5000 account is continuously compounded at 4% per year. What is it worth after 2 years?**

   _____

   _____

   _____

   _____

   _____

Review 1. What is the growth formula? _____   Calculator?

2. What's the 1st step to solve it? $1000 = Pe^{10\%(4)}$ _____   yes   no

3. What's the next step? _____

4. How does it finish? _____

5. What changes for a decay problem? $t = ne^{k(t)}$ _____

Ch 6 Ls 3 What is the Growth Formula?   341

_____ #1 #2 ____ / 7   #3 ____ / 3   R ____ / 5   Total ____ / 15   _____
Name   Checker

#1 1. What's the 1st step? $400 = 20e^{0.50(t)}$ _____

   2. How do you get rid of the e? _____

   3. How do you move the exponent? _____

   4. What happens to LN of e? _____

   5. How do you get the variable t by itself? _____

#2 1. What formula does this use?   A Rs 7000 account is continuously compounded at 8% per year. After how years is it worth Rs 7200?

   Which numbers go where?   _____

   What's the 1st step to solve it?   _____

                                      _____

                                      _____

   Estimate the answer.   _____

                                      _____

   2. What formula does this use?   A city has 8000 people, but it's depreciating at 4% each year. How long until it's 7500 people?

   Which numbers go where?   _____

   What's the 1st step to solve it?   _____

                                      _____

                                      _____

   Estimate the answer.   _____

                                      _____

342

#3 Decide how each problem uses LN.

Calculator?
yes   no

1. Tonya has 5 years until she takes a R8000 trip to Europe. How much does it take if she puts money compounded continuoucy at 5%?

_____

_____

_____

_____

_____

2. A city of 3000 has lost 100 people each year. When does the city reach 2500 people?

_____

_____

_____

_____

_____

3. A R3000 account is continuously compounded at 4% per year. What is it worth after 3 years?

_____

_____

_____

_____

_____

Review 1. What's the 1st step? **$400 = 20e^{0.50(t)}$** _____

Calculator?
yes   no

2. How do you get rid of the e? _____

3. How do you move the exponent? _____

4. What happens to LN of e? _____

5. How do you get the variable t by itself? _____

Review Problems

_____ #1 #2 ____/ 8 #3 ____/ 4 Total ____/ 12
Name

1. Natural Numbers _____
2. Continuous Growth Interest _____
3. Decay Problems _____
4. LN _____

#2 Solve these exponents.

Calculator?
yes   no

1. $e^7 (e^2)$ $e^8 (e^4)$ $e^6 (e^2)$

  _____   _____   _____

2. $\dfrac{e^4}{e^2}$ $\dfrac{e^2}{e^8}$ $\dfrac{e^6}{e^4}$

  _____   _____   _____

3. What is the formula for this?   A Rs 5000 account pays 3% per year continuenty compounded. What is it worth after 4 years?

Where does $200, 8%, and 3 go? $t =$ _____ e ─── ───

What's the first step? $t = 5000e$ ─── ───

What's the next step? $t = 5000e$ ───

What are the next steps? $t = 5000(2.71)$ ───

What is the final answer? $t = 5000($ $)$

$t =$

Continued from the last page.

2. ABC Company bought a music company for Rs 1,000,000. If declines at a rate of 20% per year. How much is it worth after 4 yrs?

#3 Decide how each problem uses LN.

Calculator?
yes   no

1. TJ has 3 years until he takes a Rs 5000 trip to Asia. How much does it take if he puts money compounded continuoucy at 4%?

2. A city of 8000 has lost 300 people each year. When does the city reach 7000 people?

3. A Rs 3000 account is continuously compounded at 5% per year. What is it worth after 4 years?

Ch 7 Ls 1 How earthquakes use logarithms.   345

_____ #1 #2 ____/ 8 #3 ____/ 6   R ____/ 6 Total ____/ 20 _____
       Name   Checker

#1 1. How do logarithms measure earthquakes? _____
    2. What key word measures earthquakes on a seismograph? _____
    3. What formula finds magnitude of an earthquake? _____
    4. What is a standard earthquake in millimeters? _____
    5. How much damage does a 3.8 quake do? _____
    6. When do quakes get dangerous? _____

#2 1. What formula does this use? _____   An earthquake's amplitude is 10,000 mm. What's the magnitude?

    Which numbers go where? $M = \log \dfrac{}{0.001}$

    What's the 1st step to solve it? _____

    Estimate the answer. _____

    _____

2. What formula does this use? _____   An earthquake's amplitude is 12,000 mm. What's the magnitude?

    Which numbers go where? $M = \log \dfrac{}{0.001}$

    What's the 1st step to solve it? _____

    Estimate the answer. _____

    _____

#3 Solve for these earthquakes.   Calculator? yes no

1. An earthquake's amplitude is 1300 mm. What's the magnitude?

   M = log $\dfrac{}{0.001}$

   M = log _____

   M = ____

   An earthquake's amplitude is 800 mm. What's the magnitude?

   M = log $\dfrac{}{0.001}$

   M = log _____

   M = ____

2. An earthquake's amplitude is 500 mm. What's the magnitude?

   M = log $\dfrac{}{0.001}$

   M = log _____

   M = ____

   An earthquake's amplitude is 900 mm. What's the magnitude?

   M = log $\dfrac{}{0.001}$

   M = log _____

   M = ____

3. An earthquake's amplitude is 400 mm. What's the magnitude?

   M = log $\dfrac{}{0.001}$

   M = log _____

   M = ____

   An earthquake's amplitude is 1500 mm. What's the magnitude?

   M = log $\dfrac{}{0.001}$

   M = log _____

   M = ____

Review
1. How do logarithms measure earthquakes? _____   Calculator? yes no
2. What key word measures earthquakes on a seismograph? _____
3. What formula finds magnitude of an earthquake? _____
4. What is a standard earthquake in millimeters? _____
5. How much damage does a 3.8 quake do? _____
6. When do quakes get dangerous? _____

Ch 7 Ls 2 How to Compare Earthquakes   347

_____ #1 #2 ____/ 8 #3 ____/ 6 R ___/ 5 Total ____/ 19 _____
                Name   Checker

#1 1. What formula compares earthquakes? _____

2. Compare a 5.9 and 5.0. Where do the numbers go? _____

3. How do you solve the exponents? _____

4. What happens to the 0.9? _____

5. What does 8 tell you about the stronger earthquake? _____

#2 1. What formula does this use?   **An earthquake of 5.0 happened and later a 5.9 occured. How much stronger was the 2nd one?**

                                          $\dfrac{10}{10}$ —— stronger quake
Which numbers go where?
                                                   —— weaker quake

What's the 1st step to solve it?   _____

Estimate the answer.   _____

                                   _____

2. Compare a 5.4 and 7.7 earthquake. $\dfrac{10}{10}$ —— stronger quake
                                              —— weaker quake

What log does this make?   ____ - ____ = ____

The stronger was ____ x stronger.   **log** ____ = ____

3. Compare a 4.0 and 5.0 earthquake. $\dfrac{10}{10}$ —— stronger quake
                                              —— weaker quake

What log does this make?   ____ - ____ = ____

The stronger was ____ x stronger.   **log** ____ = ____

#3 Compare these earthquakes.   Calculator? yes no

1. $\dfrac{10^{4.7}}{10^{4.0}}$ stronger quake / weaker quake

___ - ___ = ___

log ___ = ___

The stronger was ___ x stronger.

$\dfrac{10^{6.0}}{10^{5.2}}$ stronger quake / weaker quake

___ - ___ = ___

log ___ = ___

The stronger was ___ x stronger.

2. $\dfrac{10^{5.8}}{10^{4.7}}$ stronger quake / weaker quake

___ - ___ = ___

log ___ = ___

The stronger was ___ x stronger.

$\dfrac{10^{7.0}}{10^{5.7}}$ stronger quake / weaker quake

___ - ___ = ___

log ___ = ___

The stronger was ___ x stronger.

3. $\dfrac{10^{5.2}}{10^{3.7}}$ stronger quake / weaker quake

___ - ___ = ___

log ___ = ___

The stronger was ___ x stronger.

$\dfrac{10^{7.0}}{10^{4.9}}$ stronger quake / weaker quake

___ - ___ = ___

log ___ = ___

The stronger was ___ x stronger.

Review 1. What formula compares earthquakes? _____   Calculator? yes no

2. Compare a 5.9 and 5.0. Where do the numbers go? _____

3. How do you solve the exponents? _____

4. What happens to the 0.9? _____

5. What does 8 tell you about the stronger earthquake? _____

Ch 7 Ls 3 Logarithms and Ph   349

_____ #1 #2 ____/ 7 #3 ____/ 4 R ____/ 4 Total ____/ 15 _____
            Name   Checker

#1 1. What's the number for a neutral ph value? _____
   2. How does pH measure bases? _____
   3. What do they measure to find pH levels? _____
   4. What formula finds pH? _____

#2 1. How acidic or base are these pH evels?   **pH 3**   **pH 10**

                                                   _____   _____

   2. What does the formula subtract?   If acid rain is 5.0 and water is 7.0, how
                                         many times more acidic is the acid rain?

   How does a log use it?   ____ - ____ = ____
                                          stronger ph weaker ph

   What real number solves it?   log x = ____

   What did it find?   log ____ = ____

                                        ____ is ____ times stronger than ____.

   3. What does the formula subtract?   If acid rain is 4.2. and water is 7.0, how
                                         many times more acidic is the acid rain?

   How does a log use it?   ____ - ____ = ____
                                          stronger ph weaker ph

   What real number solves it?   log x = ____

   What did it find?   log ____ = ____

                                        ____ is ____ times stronger than ____.

#3 Compare these Ph levels to find how much stronger one is.   Calculator?
                                                                   yes   no

1. **If a lemonade has ph of 6.4 and cup of coffee ph is 5.5, what's the difference?**

   ___ - ___ = ___
   stronger ph weaker ph

   What does the formula subtract?

   How does a log use it?   log ___ = ___

   _____ is ___ times stronger than _____.

2. **If a calcium supplement has a ph of 9.2 and magnesium has 8.5, what is combination of them for a tablet?**

   ___ - ___ = ___
   stronger ph weaker ph

   How does a log use it?   log ___ = ___

   _____ is ___ times stronger than _____.

3. **If acid rain is 4.3 and water is 7.0, how many times more acidic is the acid rain?**

   ___ - ___ = ___
   stronger ph weaker ph

   How does a log use it?   log ___ = ___

   _____ is ___ times stronger than _____.

4. **If Milk of Magnesia is 10 and water is 7, how many times more basic is it?**

   ___ - ___ = ___
   stronger ph weaker ph

   How does a log use it?   log ___ = ___

   _____ is ___ times stronger than _____.

Review 1. What's the number for a neutral ph value? _____   Calculator?
                                                                                  yes   no
2. How does pH measure bases? _____

3. What do they measure to find pH levels? _____

4. What formula finds pH? _____

Ch 7 Ls 4 Logarithms and Decibels 351

_____ #1 #2 ____/ 9 #3 ____/ 3 R ____/ 7 Total ____/ 19 _____
          Name   Checker

#1 1. How is sound measured? _____

2. What are the decibels for a whisper and a jet taking off? _____

3. What unit measures decibels? _____

4. What formula changes watts/meter^2 to decibels? _____

5. Put a standard sound in the equation. _____

6. How does it compare a sound with 10^{-7} watts? _____

7. How do you solve it? _____

#2 1. What formula does this use?   A sound has amplitude of 10^{-3}
                                         How many decibels is that?

What does it subtract? $L = 10 \log \dfrac{10}{10^{-12}}$

Multiply x 10 to find decibels.   ___ - ___ = ___

$L = 10 \log 10^{\underline{}}$

$L = 10 \times \underline{} = \underline{}$

2. What formula does this use?   A sound has amplitude of 10^{-2}
                                  How many decibels is that?

What does it subtract? $L = 10 \log \dfrac{10}{10^{-12}}$

Multiply x 10 to find decibels.   ___ - ___ = ___

$L = 10 \log 10^{\underline{}}$

$L = 10 \times \underline{} = \underline{}$

#3 Compare these sound differences.   Calculator?
                                                                  yes   no

1. A sound has amplitude of 10^{-7}. $L = 10 \log \dfrac{10}{10^{-12}}$
   How many decibels is that?

   What does it subtract?   ___ - ___ = ___

   Find the answer. $L = 10 \log 10$ ———

                              $L = 10 \times$ ___ = ___

2. A sound has amplitude of 10^{-5}. $L = 10 \log \dfrac{10}{10^{-12}}$
   How many decibels is that?

   What does it subtract?   ___ - ___ = ___

   Find the answer. $L = 10 \log 10$ ———

                              $L = 10 \times$ ___ = ___

3. A sound has amplitude of 10^{-8}. $L = 10 \log \dfrac{10}{10^{-12}}$
   How many decibels is that?

   What does it subtract?   ___ - ___ = ___

   Find the answer. $L = 10 \log 10$ ———

                              $L = 10 \times$ ___ = ___

Review 1. How is sound measured? _____   Calculator?
                                                                      yes   no
2. What are the decibels for a whisper and a jet taking off? _____

3. What unit measures decibels? _____

4. What formula changes watts/meter^2 to decibels? _____

5. Put a standard sound in the equation. _____

6. How does it compare a sound with 10^{-7} watts? _____

7. How do you solve it? _____

_____ #1 ____ / 5 #2 ____ / 5   Total ____ / 10
Name

#1 1. Carbon-14 is used to determine the age of artifacts in carbon dating. It has a half-life of 5730 years. Write the exponential decay function for a 60-mg sample. Find the amount of carbon-14 remaining after 20,000 years.

Calculator?
yes   no

2. How long would it take to double the value of an account at an annual interest rate of 5% compounded continuously?

3. The GDP, of Switzerland was $600 million in 1988. Assume that GDP grows 3% each year. Use an exponential model to find which percent best describes the GDP growth that occurs over 10 years.

4. A woman at a yard store mentions that she is buying fencing for a garden that is 10 m longer than it is wide. She buys 200 m of fencing. What is the width and length of the garden?

#2 1. The noise level inside a convertible driving along the freeway with its top up is 60 dB. With the top down, the noise level is 90 dB. Find the intensity of the sound with the top up and with the top down.

Calculator?
yes   no

2. Conservation efforts have increased the endangered Florida manatee population from 1400 in 1990 to 3200 in 2001. If this growth rate continues, when might there be 10,000 manatees?

3. As a town gets smaller, the population of its high school decreases by 8% each year. The student body has 340 students now. If the town keeps getting smaller, how many years will it have about 300 students?

4. A parent raises a child's allowance by 10% each year. If the allowance is $5 now, when will it reach $10?

5. An earthquake of magnitude 7.7 occurred in 2015 in India. It was ??? times as strong as the greatest earthquake ever to hit the eastern half of the USA. Find the magnitude of the USA earthquake.

_____ #1 #2 ____/ 14 #3 #4 ____/ 8 Total ____/ 22
Name

1. Richter Scale _____
2. Amplitude _____
3. pH Level _____
4. Decibel Levels _____

#2 1. An earthquake's amplitude is 1300 mm. What's the magnitude?

$M = \log \dfrac{}{0.001}$

$M = \log \underline{}$

$M = \underline{}$

An earthquake's amplitude is 800 mm. What's the magnitude?

$M = \log \dfrac{}{0.001}$

$M = \log \underline{}$

$M = \underline{}$

Calculator? yes   no

2. An earthquake's amplitude is 500 mm. What's the magnitude?

$M = \log \dfrac{}{0.001}$

$M = \log \underline{}$

$M = \underline{}$

An earthquake's amplitude is 900 mm. What's the magnitude?

$M = \log \dfrac{}{0.001}$

$M = \log \underline{}$

$M = \underline{}$

#3 Compare these earthquakes.

1. $\dfrac{10^{4.7} \text{ stronger quake}}{10^{4.0} \text{ weaker quake}}$

____ - ____ = ____

log ____ = ____

It was ____ x stronger.

$\dfrac{10^{6.0} \text{ stronger quake}}{10^{5.2} \text{ weaker quake}}$

____ - ____ = ____

log ____ = ____

It was ____ x stronger.

2. $\dfrac{10^{5.8}}{10^{4.7}}$ stronger quake / weaker quake

___ - ___ = ___

log ___ = ___

The stronger was ___ x stronger.

$\dfrac{10^{7.0}}{10^{5.7}}$ stronger quake / weaker quake

___ - ___ = ___

log ___ = ___

The stronger was ___ x stronger.

Calculator? yes no

#4 1. If a lemonade has ph of 6.4 and cup of coffee ph is 5.5, what's the difference?
What does the formula subtract?

___ - ___ = ___
stronger ph   weaker ph

How does a log use it?

log ___ = ___

___ is ___ times stronger than ___.

2. If a calcium supplement has a ph of 9.2 and magnesium has 8.5, what is combination of them for a tablet?

___ - ___ = ___
stronger ph   weaker ph

How does a log use it?

log ___ = ___

___ is ___ times stronger than ___.

#5 1. A sound has amplitude of 10^{-7}. How many decibels is that?

$L = 10 \log \dfrac{10}{10^{-12}}$

What does it subtract?   ___ - ___ = ___

Find the answer. $L = 10 \log 10$ ‾‾‾

$L = 10 \times$ ___ = ___

2. A sound has amplitude of 10^{-5}. How many decibels is that?

$L = 10 \log \dfrac{10}{10^{-12}}$

What does it subtract?   ___ - ___ = ___

Find the answer. $L = 10 \log 10$ ‾‾‾

$L = 10 \times$ ___ = ___

Ch 8 Ls 1 Angles and Triangles 357

_____ #1 #2 ____ / 18 #3 #4 ____ /18 R ___ / 5 Total ____ /41 _____
               Name   Checker

#1 1. How does a positive angle move? _____
     2. How are the 4 quadrants numbered? _____
     3. What angles divide 90 into half and thirds? _____
     4. Name 3 steps to count basic angles. _____
        _____

     5. Find the degrees and quadrant.   80°+ 70°   20°- 120°

                              degrees _____ Quad _____   degrees _____ Quad _____

     6. Find the degrees and quadrant.   80°+ 30°   90°+ 30°

                              degrees _____ Quad _____   degrees _____ Quad _____

        7.   Estimate the
             degrees to 10s place.   a   b

                              degrees _____   degrees _____

        8.   Estimate
             the degrees.   c   d

                              degrees _____   degrees _____

#2 1. What are complementary angles? _____
     2. What are supplementary angles? _____

     3. Find complementary angles.   **50 degrees**   **40 degrees**

                                      50 + ___ = 90   40 + ___ = 90

     4. Find supplementary angles.   **70 degrees**   **110 degrees**

                                      70 + ___ = 180   110 + ___ = 180

358

#3 Add the angles. What quadrant does it end in?   Calculator? yes no

1. 90°+ 30°   40°+ 30°   210°+ 50°

    deg ____ Quad ____   deg ____ Quad ____   deg ____ Quad ____

2. -70°+ 50°   -110°- 90°   50°+ 30°

    deg ____ Quad ____   deg ____ Quad ____   deg ____ Quad ____

3. Find the complementary angles.

    50 degrees   70 degrees   35 degrees
    50 + ___ = 90   70 + ___ = 90   35 + ___ = 90

4. Find the supplementary angles.

    23 degrees   110 degrees   12 degrees
    23 + ___ = 180   110 + ___ = 180   12 + ___ = 180

#4 1.   Find the degrees using 10s.

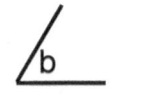

a _____   b _____   c _____

degrees ____   degrees ____   degrees ____

Calculator? yes no

2.

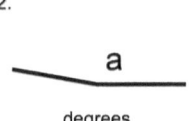 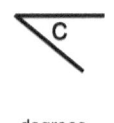

a _____   b _____   c _____

degrees ____   degrees ____   degrees ____

Review 1. How does a positive angle move? _____
2. How are the 4 quadrants numbered? _____
3. What angles divide 90 into half and thirds? _____
4. Name 3 steps to count basic angles. _____
_____
5. What angles divide 90 into half and thirds? _____

Calculator? yes no

Ch 8 Ls 2 Degees and Radians   359

_____ #1 #2 ____/10 #3 #4 ____/12 R ___/12 Total ____/34 _____
       Name   Checker

#1   1. What do radians measure? _____

2. How many degrees are in 3 4ths and a quarter pi radians? _____

3. How many radians are in half a circle? (estimated) _____

4. How many pi radians are in a full circle? _____

5. What's the 1st step to change radians to degrees? $\dfrac{2\pi}{3}$ **Radians** _____

_____

6. What's the 2nd step? _____

#2   1. Change radians to degrees. $\dfrac{\pi}{2}$ Radians
          What does it divide?

Divide 180 by 2 = ____ degrees

2. Change radians to degrees. $\dfrac{\pi}{3}$ Radians
   What does it divide?

Divide 180 by 3 = ____ degrees

3. What's the 1st step with $\dfrac{2\pi}{3}$ Radians
   a 2 step fraction?

What's the 2nd step?   **Divide 180 by 3 = ____ degrees**

Multiply ____ x ____ = ____ degrees

4. What's the 1st step with $\dfrac{2\pi}{5}$ Radians
   a 2 step fraction?

What's the 2nd step?   **Divide 180 by 5 = ____ degrees**

Multiply ____ x ____ = ____ degrees

#3 How many degrees in each radian?   Calculator? yes   no

1. $\dfrac{\pi}{6}$ Rads = ____ $\dfrac{\pi}{3}$ Rads = ____ $\dfrac{\pi}{2}$ Rads = ____

2. $\dfrac{\pi}{8}$ Rads = ____ $\dfrac{\pi}{5}$ Rads = ____ $\dfrac{\pi}{9}$ Rads = ____

3. $\dfrac{\pi}{4}$ Rads = ____ $\dfrac{\pi}{7}$ Rads = ____ $\dfrac{\pi}{10}$ Rads = ____

#4

1. $\dfrac{2\pi}{3}$ Rads = ____ $\dfrac{2\pi}{5}$ Rads = ____ $\dfrac{3\pi}{4}$ Rads = ____   Calculator? yes   no

2. $\dfrac{4\pi}{5}$ Rads = ____ $\dfrac{5\pi}{6}$ Rads = ____ $\dfrac{3\pi}{8}$ Rads = ____

3. $\dfrac{7\pi}{8}$ Rads = ____ $\dfrac{3\pi}{5}$ Rads = ____ $\dfrac{5\pi}{8}$ Rads = ____

Review

1. What do radians measure? _____   Calculator? yes   no
2. How many degrees are in 3 4ths and a quarter pi radians? _____
3. How many radians are in half a circle? (estimated) _____
4. How many pi radians are in a full circle? _____
5. What's the 1st step to change radians to degrees? $\dfrac{2\pi}{3}$ **Radians** _____

_____

6. What's the 2nd step? _____

7. $\dfrac{3\pi}{10}$ Rads = ____ $\dfrac{5\pi}{12}$ Rads = ____ $\dfrac{3\pi}{7}$ Rads = ____

8. $\dfrac{7\pi}{12}$ Rads = ____ $\dfrac{6\pi}{7}$ Rads = ____ $\dfrac{9\pi}{10}$ Rads = ____

Ch 8 Ls 3 Change degrees to radians   361

_____ #1 #2 ____/ 8 #3 ____/ 8 R ___/ 7 Total ____/ 23 _____
         Name   Checker

#1 1. What divides to change 60 degrees to radians? _____
    2. What's the 1st step to change 210 degrees to radians? _____
    3. What's the 2nd step to get radians? _____

#2 1. How do you change 30 degrees to pi radians?   **30 degrees**

                           What fraction does 6 make?   **30 x ____ is 180**

                                          30 degrees is $\boxed{}$ — radians

    2. How do you change 50 degrees to pi radians?   **50 degrees**

                          How do you simplify 50 over 180? $\dfrac{}{180}$

                                  50 degrees is $\dfrac{ \div}{180 \div}$ — = $\boxed{}$ — radians

    3. How do you change 70 degrees to pi radians?   **70 degrees**

                          How do you simplify 70 over 180? $\dfrac{}{180}$

                                  70 degrees is $\dfrac{ \div}{180 \div}$ — = $\boxed{}$ — radians

    4.   1 step, change 120 degrees to pi radians.   **120 degrees**

                                120 degrees is $\dfrac{ \div}{180 \div}$ — = $\boxed{}$ — radians

    5.   1 step, change 170 degrees to pi radians.   **170 degrees**

                                170 degrees is $\dfrac{ \div}{180 \div}$ — = $\boxed{}$ — radians

#3 How do you change degrees to pi radians?   Calculator? yes no

1. **45 degrees**   **60 degees**

    45 x ____ is 180   60 x ____ is 180

    45 degrees is $\boxed{}$ radians   60 degrees is $\boxed{}$ radians

2. **50 degrees**   **110 degrees**

    $\dfrac{ \div }{180 \div } = \boxed{}$ radians $\dfrac{ \div }{180 \div } = \boxed{}$ radians

3. **160 degrees**   **-30 degrees**

    $\dfrac{ \div }{180 \div } = \boxed{}$ radians $\dfrac{ \div }{180 \div } = \boxed{}$ radians

4. **-40 degrees**   **140 degrees**

    $\dfrac{ \div }{180 \div } = \boxed{}$ radians $\dfrac{ \div }{180 \div } = \boxed{}$ radians

Review 1. What multiplies to change 60 degrees to radians? _____   Calculator? yes no
        2. What's the 1st step to change 210 degrees to radians? _____
        3. What's the 2nd step to get radians? _____

4. **90 degrees**   **-120 degrees**

    $\dfrac{ \div }{180 \div } = \boxed{}$ radians $\dfrac{ \div }{180 \div } = \boxed{}$ radians

5. **175 degrees**   **-80 degrees**

    $\dfrac{ \div }{180 \div } = \boxed{}$ radians $\dfrac{ \div }{180 \div } = \boxed{}$ radians

Ch 8 Ls 4 Change Pi radians and degrees. 363

_____ #1 #2 ____/ 10 #3 ____/ 9 R ___/12 Total ____/31 _____
         Name   Checker

#1 1. What's the 1st step to find the angle for 1 and a half pi radians? _____
_____

2. What's the 2nd step to find the angle? _____
3. What makes the denominator for a fraction in pi radians? _____
_____

4. What's the 1st step to 1 and a 2nd pi radians? $1\frac{\pi}{2}$ Radians

A 2nd is ____ degrees.   **2nd step?**

Add 180 + ____ = ____ degrees

5. Change to degrees. How many degrees is 2 3rds pi? $1\frac{2\pi}{3}$ Radians

2 3rds is ____ degrees.   **2nd step?**

Add 180 + ____ = ____ degrees_

#2 1. What kind of an angle looks up from the ground? _____
2. What is it called when looking down from the top? _____
3. What about the 2 angles? _____

```
                                    B
                         D
              A                C
```

4. Which angle is the Angle of Depression?   Elevation?
_____   _____

5. What angles are the same?
_____

#3 Change pi radians to degrees.   Calculator?   yes   no

1. $\dfrac{3\pi}{2}$ Radians $\dfrac{4\pi}{3}$ Radians $\dfrac{7\pi}{6}$ Radians

   180 + ___ = ___ deg   180 + ___ = ___ deg   180 + ___ = ___ deg

2. $\dfrac{5\pi}{3}$ Radians $\dfrac{7\pi}{4}$ Radians $\dfrac{6\pi}{5}$ Radians

   180 + ___ = ___ deg   180 + ___ = ___ deg   180 + ___ = ___ deg

3. $\dfrac{9\pi}{8}$ Radians $\dfrac{5\pi}{4}$ Radians $\dfrac{11\pi}{6}$ Radians

   180 + ___ = ___ deg   180 + ___ = ___ deg   180 + ___ = ___ deg

Review 1. What's the 1st step to find the angle for 1 and a half pi radians? _____   Calculator?   yes   no

_____

2. What's the 2nd step to find the angle? _____

3. What makes the denominator for a fraction in pi radians? _____

_____

4. What kind of an angle looks up from the ground? _____

5. What is it called when looking down from the top? _____

6. What about the 2 angles? _____

7. $\dfrac{11\pi}{10}$ Radians $\dfrac{13\pi}{10}$ Radians $\dfrac{19\pi}{18}$ Radians

   180 + ___ = ___ deg   180 + ___ = ___ deg   180 + ___ = ___ deg

8. $\dfrac{21\pi}{20}$ Radians $\dfrac{23\pi}{18}$ Radians $\dfrac{39\pi}{20}$ Radians

   180 + ___ = ___ deg   180 + ___ = ___ deg   180 + ___ = ___ deg

Ch 8 Ls 5 Degrees, Radians, and Revolutions   365

_____ #1 #2 ____/ 9 #3 ____/12 R ____/12 Total ____/31 _____
          Name   Checker

#1 1. How many radians are in 1, 2, and 3 revolutions? 1 is _____ 2 is _____ 3 is _____

  2. Name 3 steps to change radians to degrees. **1. Make a** _____.

   2. Count whole _____. **3. Add the** _____.

  3. How many radians are in 420 degrees? _____

#2 1. 1st step to change degrees into radians?   **540 degrees**

                          _____ whole pi is _____ degrees   What is left?

                          Subtract makes _____ degrees   Radian answer?

                                    _____ radians

  2. 1st step to change degrees into radians?   **720 degrees**

                          _____ whole pi is _____ degrees   What is left?

                                    _____ radians   Radian answer?

  3. What's the 1st step to change radians to degrees? $\dfrac{5\pi}{2}$

                     _____ whole circles is _____ degrees   What's the 2nd step?

   $\dfrac{\pi}{2}$ rad is _____ degrees are left $\dfrac{5\pi}{2}$ is _____ degrees

  4. What's the 1st step to change radians to degrees? $\dfrac{7\pi}{3}$

                     _____ whole circles is _____ degrees

   $\dfrac{\pi}{3}$ rad is _____ degrees are left $\dfrac{7\pi}{3}$ is _____ degrees

#3 Change radians to be degrees.   Calculator? yes no

1. $\dfrac{7\pi}{2}$ Rds = _____ deg $\dfrac{7\pi}{3}$ Rds = _____ deg $\dfrac{9\pi}{4}$ Rds = _____ deg

2. $\dfrac{10\pi}{2}$ Rds = _____ deg $\dfrac{11\pi}{4}$ Rds = _____ deg $\dfrac{11\pi}{3}$ Rds = _____ deg

3. $\dfrac{12\pi}{5}$ Rds = _____ deg $\dfrac{5\pi}{4}$ Rds = _____ deg $\dfrac{5\pi}{2}$ Rds = _____ deg

4. $\dfrac{8\pi}{3}$ Rds = _____ deg $\dfrac{7\pi}{4}$ Rds = _____ deg $\dfrac{5\pi}{3}$ Rds = _____ deg

Review 1. How many radians are in 1, 2, and 3 revolutions? 1 is _____ 2 is _____ 3 is _____   Calculator? yes no

2. Name 3 steps to change radians to degrees. **1. Make a** _____

   2. Count whole _____ . **3. Add the** _____

3. How many radians are in 420 degrees? _____

Change degrees using revolutions.

4.
540 degrees	480 degrees	400 degrees
_____ whole circles	_____ whole circles	_____ whole circles
_____ is _____ radians	_____ is _____ radians	_____ is _____ radians

5.
600 degrees	370 degrees	450 degrees
_____ whole circles	_____ whole circles	_____ whole circles
_____ is _____ radians	_____ is _____ radians	_____ is _____ radians

6.
-350 degrees	-390 degrees	-540 degrees
_____ whole circles	_____ whole circles	_____ whole circles
_____ is _____ radians	_____ is _____ radians	_____ is _____ radians

Review Problems 367

_____ #1 ____/ 6 #2 ____/ 5 Total ____/ 11
Name

#1 1. A square has four right angle corners. Give an example of another shape that has four right angle corners.

Calculator?
yes   no

2. Marshall lives 2300 meters from school and 1500 meters from the pharmacy. The school, pharmacy, and his home are linear. What is the total distance from the pharmacy to the school?

   pharmacy   Marshall   school

3. Mr K is driving down A Street heading east. He takes a right onto B Street. What type of angle did he have to turn his car through? What is the angle measure is turning his car when he takes the right turn?

   55/A   B

4. A sign painter is painting a large "X". What are the measures of angles 1, 2, and 3?

   1 2 3
      65

   1 is _____
   2 is _____
   3 is _____

5 Amav was walking home when he looked at the letters on a street sign and noticed how many are made up of angles. The sign he looked at was KLINE ST. Which letter(s) on the sign have an obtuse angle? What other letters in the alphabet have a rigth angle?

   obtuse _____
   right _____

6 A hiking trail is 30 kilometers long. Park organizers want to build 5 rest stops for hikers with one on each end of the trail and the other 3 spaced evenly between. How much distance will separate successive rest stops?

   stops _____

#2 Story Problems.

Calculator?
yes   no

1. A 1 meter pendulum swings through an angle of 24 deg. Find the length of the arch in centimeters through which the tip of the pendulum swings.

   _____
   _____
   _____

2. The area of a circle with radius 10 cm is 20 square cm. Find the measure of the central angle of the sector in degrees.

   _____
   _____
   _____

3. A ferris wheel's radius is 20 meters. You measure the time it takes for one revolution is 70 seconds. What is the linear speed in meters/sec?

   _____
   _____
   _____

4. The outer diameter of the wheels on a bicycle is 54 centimeters. If the wheels are turning at a rate of 200 rpm. Find the linear speed of the bike in meters per minute.

   _____
   _____
   _____

5. The windshield wiper of a car is 40 centimeters long. How many centimeters will the tip of the wiper trace out in?

   _____
   _____
   _____

6. A wheel is rotating at 30 rpm. Find the angular speed in radians per second.

   _____
   _____
   _____

_____ #1 - #3 ____/ 49 Back ____/ 11 Total ____/ 60
Name

1. Pi _____
2. Radian _____
3. Revolution _____
4. Angles of Elevation _____
5. Angles of Depression _____

#2 Add the angles. What quadrant does it end in?   Calculator?
yes   no

1. 90°+ 30°   40°+ 30°   210°+ 50°

   deg ____ Quad ____   deg ____ Quad ____   deg ____ Quad ____

2. -70°+ 50°   -110°- 90°   50°+ 30°

   deg ____ Quad ____   deg ____ Quad ____   deg ____ Quad ____

3.
Find the degrees
using 30s.

   a   b   c

   degrees ____   degrees ____   degrees ____

4.
   a   b   c

   degrees ____   degrees ____   degrees ____

#3 How many degrees in each radian?   Calculator?
yes   no

1. $\frac{\pi}{6}$ Rads = ____ $\frac{\pi}{3}$ Rads = ____ $\frac{\pi}{2}$ Rads = ____

2. $\frac{\pi}{12}$ Rads = ____ $\frac{\pi}{5}$ Rads = ____ $\frac{\pi}{9}$ Rads = ____

3. $\frac{\pi}{4}$ Rads = ____ $\frac{\pi}{15}$ Rads = ____ $\frac{\pi}{10}$ Rads = ____

#4 How do you change degrees to pi radians?   Calculator? yes no

1. **50 degrees** $\dfrac{ \div }{180 \div } = \dfrac{\square}{\square}$ radians   **110 degrees** $\dfrac{ \div }{180 \div } = \dfrac{\square}{\square}$ radians

2. **160 degrees** $\dfrac{ \div }{180 \div } = \dfrac{\square}{\square}$ radians   **-30 degrees** $\dfrac{ \div }{180 \div } = \dfrac{\square}{\square}$ radians

3. **-40 degrees** $\dfrac{ \div }{180 \div } = \dfrac{\square}{\square}$ radians   **140 degrees** $\dfrac{ \div }{180 \div } = \dfrac{\square}{\square}$ radians

#5 Change pi radians to degrees.   Calculator? yes no

1. $\dfrac{3\pi}{2}$ Radians $\dfrac{4\pi}{3}$ Radians $\dfrac{7\pi}{6}$ Radians
   180 + ___ = ___ deg   180 + ___ = ___ deg   180 + ___ = ___ deg

2. $\dfrac{5\pi}{3}$ Radians $\dfrac{7\pi}{4}$ Radians $\dfrac{6\pi}{5}$ Radians
   180 + ___ = ___ deg   180 + ___ = ___ deg   180 + ___ = ___ deg

#6 Change radians to be degrees.   Calculator? yes no

1. $\dfrac{7\pi}{2}$ Rds = ___ deg $\dfrac{7\pi}{3}$ Rds = ___ deg $\dfrac{9\pi}{4}$ Rds = ___ deg

2. $\dfrac{10\pi}{2}$ Rds = ___ deg $\dfrac{11\pi}{4}$ Rds = ___ deg $\dfrac{11\pi}{3}$ Rds = ___ deg

3. $\dfrac{12\pi}{5}$ Rds = ___ deg $\dfrac{5\pi}{4}$ Rds = ___ deg $\dfrac{5\pi}{2}$ Rds = ___ deg

Ch 9 Ls 1 How Trigonometry uses Triangles. 371

_____ #1 #2 ____ / 8   #3 ____ / 9   R ___ / 5   Total ____ / 22   _____
           Name   Checker

#1 1. What formula finds the length of a side of a right triangle? _____

2. How do you label the angles? _____

3. What is a unit circle? _____

4. Why is the radius so important? _____

5. How many degrees are in every triangle? _____

#2 1. What equation solves this?

Solve the next step. _____

What's the answer? _____

2. What equation solves this?

Solve the next step. _____

What's the answer? _____

3. What equation solves this?

Solve the next step. _____

What's the answer? _____

#3 Use Pythagorean to solve these.   Calculator?
                                                yes   no

1. 8 / a / 3   c / 1 / 5   7 / a / 4

_____   _____   _____

_____   _____   _____

_____   _____   _____

Not all triangles
are drawn to scale.

2. c / 1 / 5   6 / 1 / b   5 / a / 3

_____   _____   _____

_____   _____   _____

_____   _____   _____

3. 7 / a / 5   c / 2 / 5   6 / a / 3

_____   _____   _____

_____   _____   _____

_____   _____   _____

Review 1. What are the labels for the sides of a right triangle? _____   Calculator?
                                                                                      yes   no
2. Where do the angles go? _____

3. What formula finds a side of a right triangle? _____

4. What is a unit circle? _____

5. How many degrees are in every triangle? _____

Ch 9 Ls 2 What is the sin ratio? 373

_____ #1 #2 ____/ 9 #3 ____/12 R ____/ 8 Total ____/ 29 _____
         Name   Checker

#1 1. What is the sin ratio? _____

2. How can you estimate sine of 6 to 36 degrees? Each 6 degrees is _____

3. How do you find other numbers from a sin ratio? _____

4. What's the Sine of 0 degrees? _____

#2 1. How many 10ths are these?   **18 degrees**   **30 degrees**

3 x 6 Ratio ____10ths   5 x 6 Ratio ____10ths

2. How many 10ths are these?   **12 degrees**   **24 degrees**

2 x 6 Ratio ____10ths   4 x 6 Ratio ____10ths

3. What's the sin ratio and a side?   10 yd 12° a

10 yd 12°   ratio _____
              _____ yd

4. Find the sin ratio and a side.   30 yd 24° a

30 yd 24°   ratio _____
              _____ yd

5. What's the sin ratio and a sin?   20 yd 18° a

20 yd 18°   ratio _____
              _____ yd

#3 Find the sin ratio and solve for the side.   Calculator? yes no

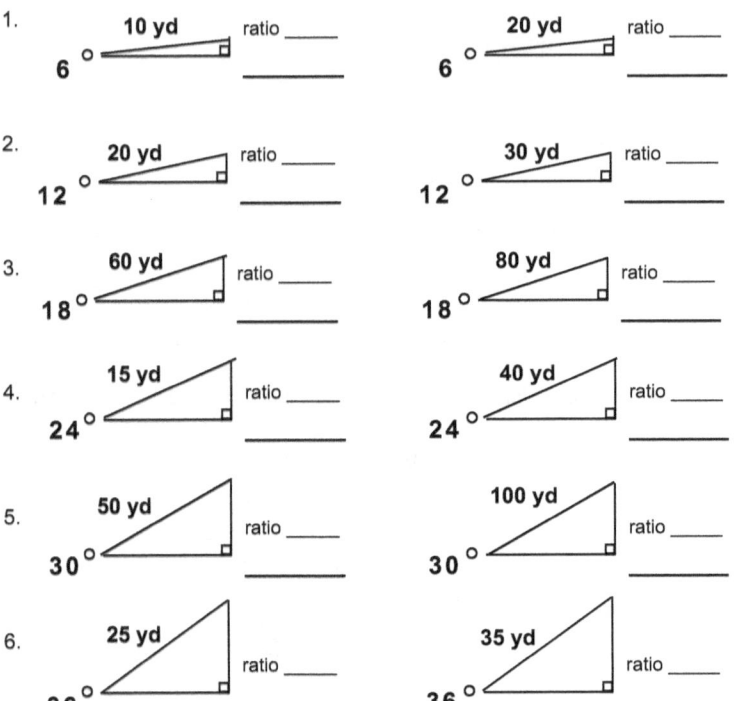

Review 1. What is the sin ratio? _____   Calculator?
2. How can you estimate sine of 6 to 36 degrees? Each 6 degrees is _____   yes no
3. How do you find other numbers from a sin ratio? _____
4. What's the Sine of 0 degrees? _____

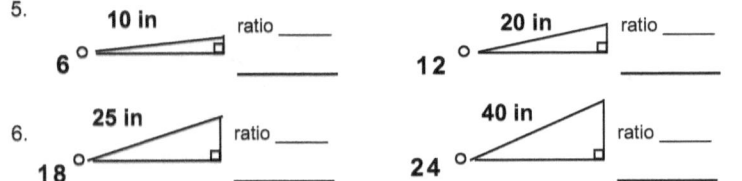

Ch 9 Ls 3 Find the Sin Ratios 45 and over. 375

_____ #1 #2 ____/ 6 #3 #4 ____/ 10 R ___/ 3 Total ____/ 19 _____
         Name   Checker

#1 1. What is the sin ratio for 45 degrees? _____

    2. What's the sin for 90 degrees? _____

    3. What's the sin ratio for 53 and 64 degrees? 53 is ____10ths   64 is ____10ths

#2 1. What's the sin ratio and a side?

        10 yd   a
        45°

        10 yd
                     ratio _____
        45°
                     _____ yd

    2. Find the sin ratio and a side.

        10 yd   a
        53°

        10 yd
                     ratio _____
        53°
                     _____ yd

    3. Find the sin ratio and a side.

        10 yd   a
        64°

        10 yd
                     ratio _____
        64°
                     _____ yd

#3 Find the sin ratio and solve for the side.   Calculator? yes no

1.
   20 yd   45°   ratio _____   _____
   30 yd   54°   ratio _____   _____

2.
   40 yd   63°   ratio _____   _____
   50 yd   90°   ratio _____   _____

#4 Answer these story problems mentally.   Calculator? yes no

1. A ladder is leaning at 63 degrees. It's 4 m long. How high will it reach?

   A ladder is leaning at 54 degrees. It's 8 m long. How high will it reach?

2. A ramp has an angle of 6 degrees. The ramp is 6 m long. How tall is it?

   Another ramp has a distance that is 12 degrees. It's 20 m long. How tall is it?

3. A pyramid has bases that are 15 m long. It's height is 45 degrees. How tall is it?

   A pyramid has bases that are 12 m long. It's height is 36 degrees. How tall is it?

Review 1. What is the sin ratio for 45 degrees? _____   Calculator? yes no

2. What's the sin for 90 degrees? _____

3. What's the sin ratio for 53 and 64 degrees? _____

Ch 9 Ls 4 How triangles make roots. 377

_____ #1 #2 ____/ 7 #3 ____/ 6 R ____/ 4 Total ____/ 17 _____
            Name   Checker

#1 1. How do triangles make square roots? _____

2. If a hypotenuse is 2, how does it's triangle fit a unit circle? _____

_____

3. How do you find the 3rd angle with a right triangle? _____

4. How do you find the 3rd side? _____

#2 1. Find the sin ratio and solve for the side.
         What equation finds the 3rd side?   5 in ⟋⎹ 2 in
                                                      b

                    Solve the 1st step.   __^{2} + __^{2} = __^{2}

                    Solve the 2nd step. Estimate it.   _____

                                                      _____

2. What equation finds the 3rd side?   7 ft ⟋⎹ 1 ft
                                                b

                    Solve the 1st step.   __^{2} + __^{2} = __^{2}

                    Solve the 2nd step. Estimate it.   _____

                                                      _____

                                                  6 yd ⟋⎹ 3 yd
3. What equation finds the 3rd side?   b

                    Solve the 1st step.   __^{2} + __^{2} = __^{2}

                    Solve the 2nd step. Estimate it.   _____

                                                      _____

#3 Solve for the root side.   Calculator? yes no

1. 50 ft ╱| 20 ft   40 in ╱| 15 in
         b   b

   __² + __² = __²   __² + __² = __²

   _____   _____
   _____   _____

2. 16 yd ╱| 5 yd   100 ft ╱| 40 ft
          b   b

   __² + __² = __²   __² + __² = __²

   _____   _____
   _____   _____

3. 60 cm ╱| 30 cm   8 meters ╱| 3 meters
          b   b

   __² + __² = __²   __² + __² = __²

   _____   _____
   _____   _____

Review 1. How do triangles make square roots? _____   Calculator?
2. If a hypotenuse is 2, how does it's triangle fit a unit circle? _____   yes no

_____

3. How do 2 sides of a triangle make a sin ratio? _____
4. How do you find the 3rd side? _____

_____ #1 ____/7 #2 ____/ 15 Total ____/39
Name

Calculator?
yes   no

#1 1. A 20 meter flagpole casts a shadow of 10 meters. Find the angle of elevation of the sun.

Make an equation first.

2. A flat 5 meter plank rests with one end on the ground and the other end upon a 1 meter ledge. How far from the base of the ledge is the far end of the plank?

3. A submarine travels at a depth of 40 meters dives at an angle of 10° with respect to a line parallel to the water's surface. It travels a horizontal distance of 900 meters during the dive. What is the depth of the submersible after the dive?

4. Ojas buys a tent that has a center pole 2 meters high. If the sides of the tent are supposed to make a 45° angle with the ground, how wide is the tent?

5. A fire department's longest ladder is 30 m long. The rules state that they can use it for rescues up to 25 meters off the ground. What is the maximum safe angle of elevation for the rescue ladder?

6. A 10 meter pole has a support wire that runs from its top to the ground with an angle of depression of 60°. The wire is 15 m long. How far is the base to the base of the wire?

7. You're flying a kite, and it gets caught at the top of the tree. You've let out all 60 meters of string for the kite, and the angle that the string makes with the ground is 54 degrees. You wonder. "How tall is that tree?"

#2

1. A swimming pool is 40 meters long. The bottom of the pool is slanted so that the water depth is 1.0 meter at the shallow end and 5 meters at the deep end. Find the angle of depression for the pool.

Calculator?
yes   no

2. A tower is 500 meters tall. If you are on level ground exactly 1 kilometer from the base of the tower, what is your angle of elevation as you look at the tower?

3. At a point 100 meters from the base of a smokestack to the top is 50°. How tall is the smokestack?

4. An observer is at point A as she watches a rocket carrying a satellite. The observer is 600 meters from the launch. After 30 seconds, she judges the angle to be 60 degrees. How far away is the rocket now?

5. Mr W is walking to his office building which he knows is 30 meters high. The angle to the top of the building from his current location is 10°. How far is he away from the building?

Review Problems **381**

_____ #1 #2 #3 ____ / 16   #4 #5 #6 ____ / 12   Total ____ / 28
                Name

1. **Unit Circle** _____

2. **Sin** _____

3. **Sin Angles** _____

4. **Sin Over 45 d** _____

#2 Use Pythagorean to solve these.   Calculator?
                                                    yes   no

1.   8 / a / 3   c / 1 / 5   7 / a / 4

Not all triangles
are drawn to scale.

_____   _____   _____
_____   _____   _____
_____   _____   _____

2.   c / 3 / 8   6 / 2 / b   9 / a / 7

_____   _____   _____
_____   _____   _____
_____   _____   _____

#3 Find the sin ratio and solve for the side.   Calculator?
                                                    yes   no

1.   15 yd / 6°   ratio ____   40 yd / 24°   ratio ____
                   ____   ____

2.   20 yd / 12°   ratio ____   100 yd / 30° ratio ____
                   ____   ____

3.   60 yd / 18°   ratio ____   35 yd / 36°   ratio ____
                   ____   ____

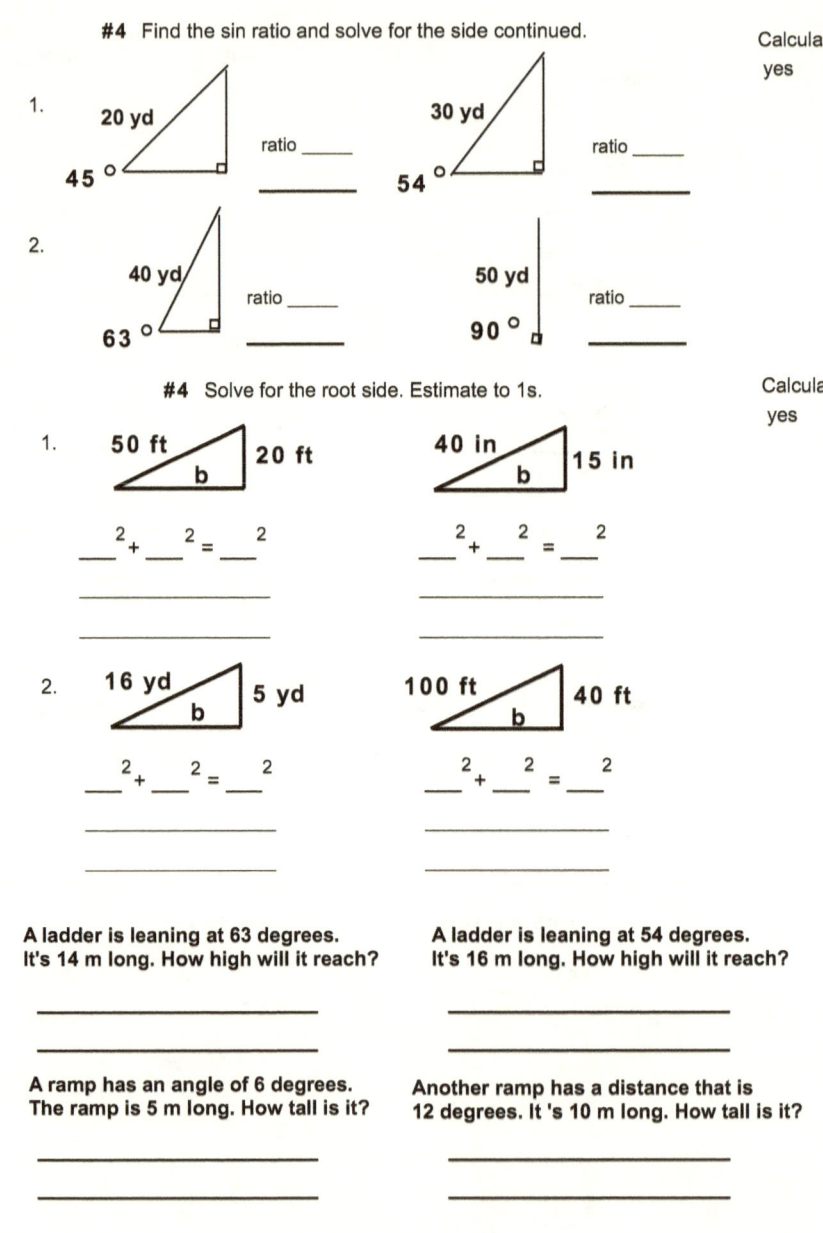

#4 Find the sin ratio and solve for the side continued.   Calculator? yes no

1. 20 yd, 45°, ratio ____ ____
   30 yd, 54°, ratio ____ ____

2. 40 yd, 63°, ratio ____ ____
   50 yd, 90°, ratio ____ ____

#4 Solve for the root side. Estimate to 1s.   Calculator? yes no

1. 50 ft, 20 ft, b
   $\_\_^2 + \_\_^2 = \_\_^2$

   40 in, 15 in, b
   $\_\_^2 + \_\_^2 = \_\_^2$

2. 16 yd, 5 yd, b
   $\_\_^2 + \_\_^2 = \_\_^2$

   100 ft, 40 ft, b
   $\_\_^2 + \_\_^2 = \_\_^2$

Story Problems

#5

1. A ladder is leaning at 63 degrees. It's 14 m long. How high will it reach?

   A ladder is leaning at 54 degrees. It's 16 m long. How high will it reach?

2. A ramp has an angle of 6 degrees. The ramp is 5 m long. How tall is it?

   Another ramp has a distance that is 12 degrees. It's 10 m long. How tall is it?

Ch 10 Ls 1 Cosine Ratios 383

_____ #1 #2 ____/ 9 #3 ____/10 R ___/ 6 Total ____/ 25 _____
         Name   Checker

#1 1. What ratio does Cosine find? _____

2. What are these cosine angles?   6 deg _____   12 deg _____   18 deg _____
3. Find these cosine angles.   24 deg _____   30 deg _____   36 deg _____
4. What are these cosine angles?   45 deg _____   53 deg _____   64 deg _____

#2 1. What's the cosine angle for sin?   **84 degrees**

                                  _____ degrees is _____ cos ratio

    2. What's the cosine angle for sin?   **78 degrees**

                                  _____ degrees is _____ cos ratio

    3. What's the cosine angle for sin?   **66 degrees**

                                  _____ degrees is _____ cos ratio

    4. What's the cosine ratio and a side?   10 yd   B
                                          12°⌐_____⌐ a

                                              10 yd   ------ratio _____
                                          12°⌐_____⌐
                                                              _____ yd

    5. Find the sin ratio and a side.   10 yd   B
                                          18°⌐_____⌐ a

                                              10 yd   ------ratio _____
                                          18°⌐_____⌐
                                                              _____ yd

#3 Find the cosine ratio and solve for the side.   Calculator? yes no

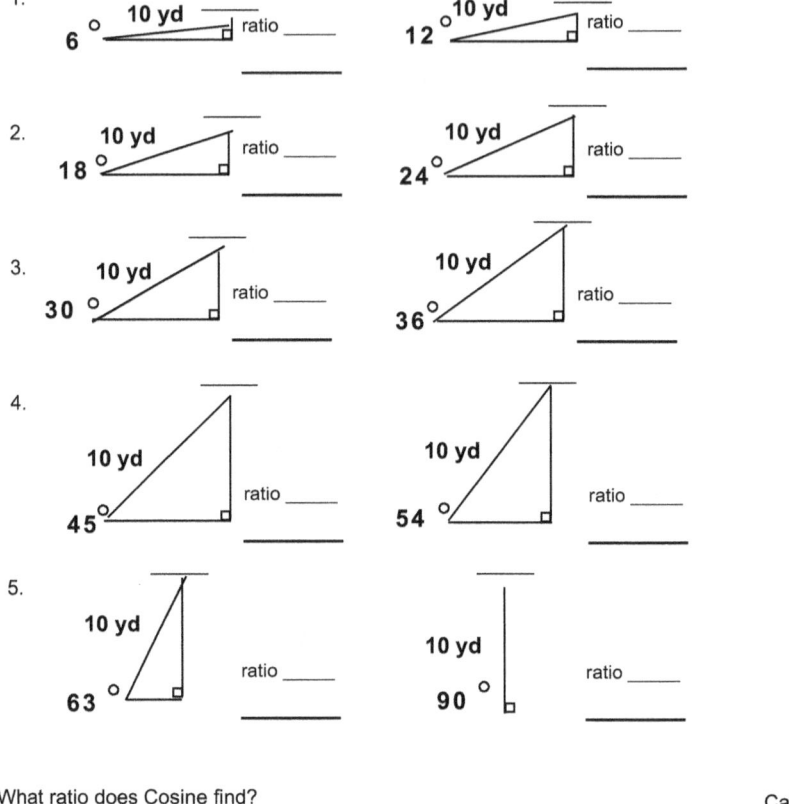

Review 1. What ratio does Cosine find? _____ Calculator?
2. What are these cos angles? 6 deg _____   12 deg _____   18 deg _____   yes no
3. Find these cos angles.   24 deg _____   30 deg _____   36 deg _____
4. What are these cos angles? 45 deg _____   53 deg _____   64 deg _____

5.
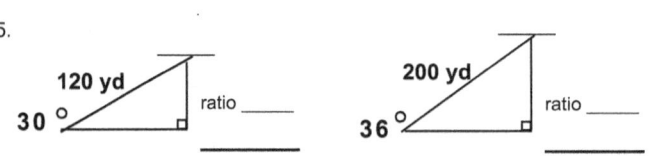

Ch 10 Ls 2 Find the cosine ratios. 385

_____ #1 #2 ____/ 6   #3 ____/ 8   R ____/ 2   Total ____/ 16 _____
         Name   Checker

#1 1. How does sine find the cosine fraction? _____
        2. What ratio does 0.4 to 0.9 stand for? _____

#2 1. What's the sin ratio and a side?   10 yd   78° a
                                              C

                                              10 yd   78° ratio ____
                                                                              ____ yd
                                              ____

        2. What's the sin ratio and a side?   10 yd 84° a
                                              C

                                              10 yd   84° ratio ____
                                                                              ____ yd
                                              ____

        3. What's the sin ratio and a side?   10 yd 72° a
                                              C

                                              10 yd   72° ratio ____
                                                                              ____ yd
                                              ____

        4. What's the sin ratio and a side?   10 yd 66° a
                                              C

                                              10 yd   66° ratio ____
                                                                              ____ yd
                                              ____

#3 Find the cosine ratio and solve for the side.   Calculator? yes no

1. 72° 10 yd   ratio _____

    84° 10 yd   ratio _____

2. 60° 10 yd   ratio _____

    66° 10 yd   ratio _____

3. 36° 10 yd   ratio _____

    45° 10 yd   ratio _____

4. Answer these story problems mentally.

1. A ladder is leaning at 18 deg cos. It's 10 m long. How high will it reach?

   A ladder is leaning at 24 deg cos. It's 10 m long. How high will it reach?

Review 1. How does sine find the cosine fraction? _____   Calculator?
        2. What ratio does 0.4 to 0.9 stand for? _____   yes no

Ch 10 Ls 3 Tangent Ratios   387

_____ #1 #2 ____/ 15 #3 ____/10 R ___/ 7 Total ____/ 32 _____
　　　Name　　　　　　　　　　　　　　　　　　　　　　　　　　　　　　　　Checker

#1 1. Which sides of a triangle ratio make up tangent? _____
　　　2. How do you estimate the tangent ratios of 6 to 31 degrees? _____
　　　3. What angles find the ratio for 8 and 9 tenths?　　8 tenths ____　9 tenths ____
　　　4. How many 10ths are these?　　**6 degrees**　　　**11 degrees**

　　　　　　　　　　　　　　　　　1 x 5 + 1 is ___ 10ths　　2 x 5 + 1 is ___ 10ths

　　　5. How many 10ths are these?　　**16 degrees**　　　**21 degrees**

　　　　　　　　　　　　　　　　　3 x 5 + 1 is ___ 10ths　　4 x 5 + 1 is ___ 10ths

　　　6. How many 10ths are these?　　**26 degrees**　　　**31 degrees**

　　　　　　　　　　　　　　　　　5 x 5 + 1 is ___ 10ths　　6 x 5 + 1 is ___ 10ths

　　　7. How many 10ths are these?　　**36 degrees**　　　**39 degrees**

　　　　　　　　　　　　　　　　　7 x 5 + 1 is ___ 10ths　　　　　___ 10ths

　　　8. How many 10ths are these?　　**42 degrees**　　　**45 degrees**

　　　　　　　　　　　　　　　　　　___ 10ths　　　　　　___ 10ths

#2 1. What's the tan ratio and a side?
　　　　　　　　　　　　　　　　6° ⟍____a
　　　　　　　　　　　　　　　　　10 yd

　　　　　　　　　　　　　　　　6° ⟍____　ratio ____
　　　　　　　　　　　　　　　　　10 yd
　　　　　　　　　　　　　　　　　　　　　　_____ yd

　　　2. What's the tan ratio and a side?
　　　　　　　　　　　　　　　　11° ⟍____a
　　　　　　　　　　　　　　　　　10 yd

　　　　　　　　　　　　　　　　　　　　　　ratio ____
　　　　　　　　　　　　　　　　11° ⟍____
　　　　　　　　　　　　　　　　　10 yd　　_____ yd

#3 Find the tangent ratio and solve for the side.   Calculator? yes no

1. 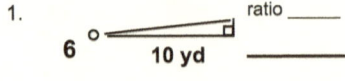 ratio ____ ____   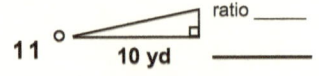 ratio ____ ____
   6° 10 yd   11° 10 yd

2. 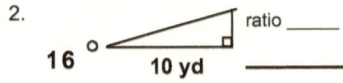 ratio ____ ____   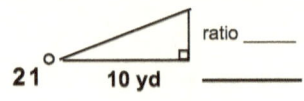 ratio ____ ____
   16° 10 yd   21° 10 yd

3. 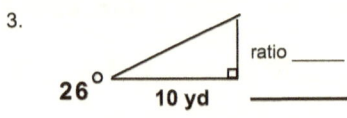 ratio ____ ____   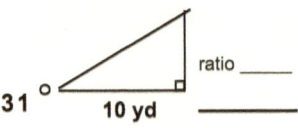 ratio ____ ____
   26° 10 yd   31° 10 yd

4. 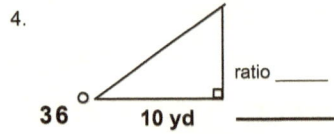 ratio ____ ____   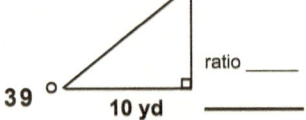 ratio ____ ____
   36° 10 yd   39° 10 yd

5. 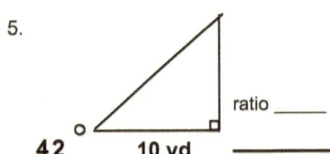 ratio ____ ____   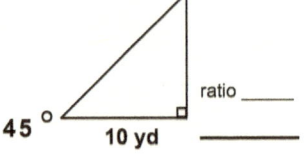 ratio ____ ____
   42° 10 yd   45° 10 yd

Review 1. Which sides of a triangle ratio make up tangent? _____   Calculator? yes no
2. How do you estimate the tangent ratios of 6 to 31 degrees? _____
3. What angles find the ratio for 8 and 9 tenths?   8 tenths ____   9 tenths ____

Use the tangent ratio to solve it.

4. 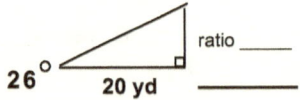 ratio ____ ____   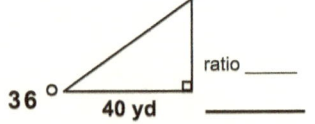 ratio ____ ____
   26° 20 yd   36° 40 yd

5. 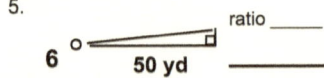 ratio ____ ____   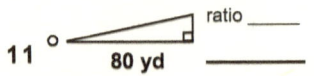 ratio ____ ____
   6° 50 yd   11° 80 yd

Ch 10 Ls 4 Tangent Ratios over 45   389

_____ #1 #2 ____/ 7 #3 ____/ 6 R ___/ 5 Total ____/ 18 _____
           Name   Checker

#1 1. Which tangent ratio has equal sides? _____

2. What are the tangent ratios?   63 deg _____   72 deg _____   76 deg _____

3. What are the tangent ratios?   84 deg _____   87 deg _____   89 deg _____

4. What's the tan ratio and a side?

63° 10 yd a

72° 10 yd a

63° 10 yd   ratio _____   _____ yd

72° 10 yd   ratio _____   _____ yd

#2 1. What is the pattern for negative numbers? _____

_____

2. What is the rate for negative 6 degrees?

_____ rate

3. What is the rate for negative 45 degrees?

_____ rate

4. What is the rate for negative 72 degrees?

_____ rate

#3 1.

Find the tangent ratio.

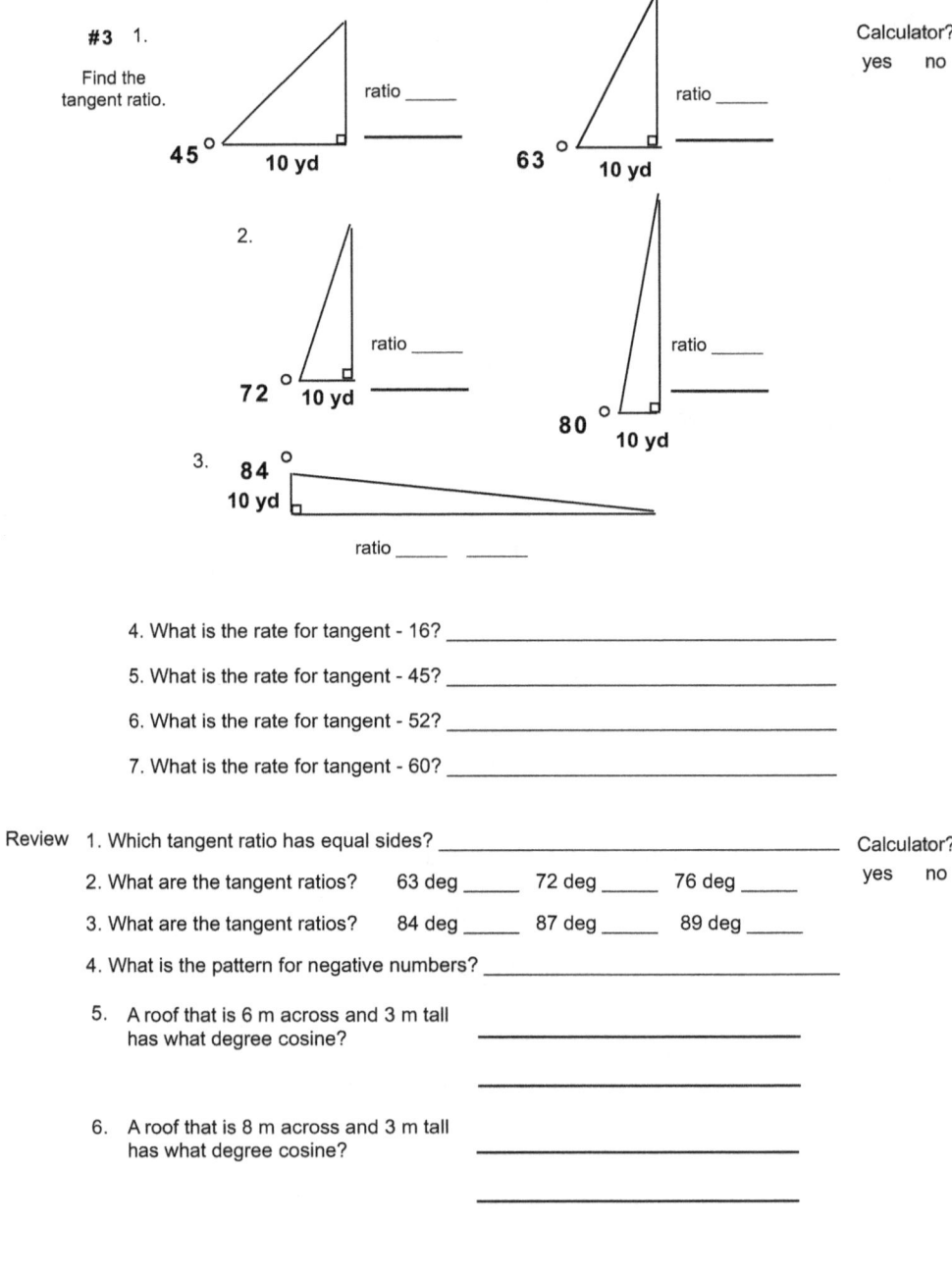

Calculator?
yes   no

45° 10 yd   ratio _____

63° 10 yd   ratio _____

2.

72° 10 yd   ratio _____

80° 10 yd   ratio _____

3.

84° 10 yd   ratio _____ _____

4. What is the rate for tangent - 16? _____

5. What is the rate for tangent - 45? _____

6. What is the rate for tangent - 52? _____

7. What is the rate for tangent - 60? _____

Review 1. Which tangent ratio has equal sides? _____   Calculator?
2. What are the tangent ratios?   63 deg _____ 72 deg _____ 76 deg _____   yes   no
3. What are the tangent ratios?   84 deg _____ 87 deg _____ 89 deg _____
4. What is the pattern for negative numbers? _____

5. A roof that is 6 m across and 3 m tall has what degree cosine?   _____
_____

6. A roof that is 8 m across and 3 m tall has what degree cosine?   _____
_____

Review Problems 391

_____ #1 ____/ 5 #1 ____/ 5 Total ____/ 10
                Name

Calculator?
yes   no

#1 1. A toy ladder is set against a 40 cm tall. If the   _____
       base of the ladder is 10 cm away from the
       base, what angle of elevation does the ladder   _____
       form?

Make an equation
before solving.   _____

    2. A submarine is 40 meters away from an under-
       water bridge it needs to clear. It maintains a   _____
       diving angle of 18 degrees. How far is the sub
       under water when it clears the bridge?   _____

                                                             _____

                                                             _____

    3. A surveyor wants to find the distance between
       peaks A and B. He finds point C, 100 m from   _____
       peak?
                                                             _____

                                                             _____

                                                             _____

    4. A person flying a kite has released 200 m of   _____
       string. The string makes an angle of 24° with
       the ground. How high is the kite? How far   _____
       away is the kite horizontally?

                                                             _____

                                                             _____

    5. A ladder is leaning up against a house. The base   _____
       of the ladder is 1 m away from the building. The
       ladder is 8 m tall. How high up the building does   _____
       the ladder reach?

                                                             _____

                                                             _____

#4

1. Mr T has a ladder and he stands 6 meters from the base of the tower and looks up at his wife, the angle of elevation to her window is 45 deg. How long does the ladder have to be?

Calculator? yes   no

2. A wheelchair ramp is 6 meters long. It rises up at a half meter. What is its angle of inclination? (Estimate it.)

3. A 6 meter ladder rests against the side of a wall. The base of the ladder is 2 meters from the base of the wall. Determine the measure of the angle between the ladder and the ground.

4. From a horizontal distance of 80 meters, the angle of elevation to the top of a flagpole is 24°. Calculate the height of the flagpole.

5. The angle of elevation of the sun is 60° when a tree casts a shadow 10 meters long. How tall is the tree?

Review Problems 393

_____ #1 #2 #3 ____/ 10 #4 #5 ____/ 13 Total ____/ 23
            Name

1. Cosine _____

2. Tangent _____

#2 Find the cosine ratio and solve for the side.   Calculator?
                                                                       yes   no
1.
      6° 15 ft   ratio _____   30 ft   ratio _____
                                   24°
      _____   _____

2.   20 ft
   18°   ratio _____   12° 10 ft   ratio _____
      _____   _____

3.   40 ft
   30°   ratio _____   120 ft
      _____   54°   ratio _____
                                       _____

4.   100 ft
                                    50 ft
   45°   ratio _____   36°   ratio _____
      _____   _____

#3 Find the cosine ratio and solve for the side.   Calculator?
                                                                       yes   no
1.   72°   84°
     4 m   6 m   ratio _____
             ratio _____
     _____   _____   _____

                                       66°
2.   60°   20 m
     10 m   ratio _____
             ratio _____
     _____   _____   _____

#4 Find the tangent ratio and solve for the side.   Calculator? yes no

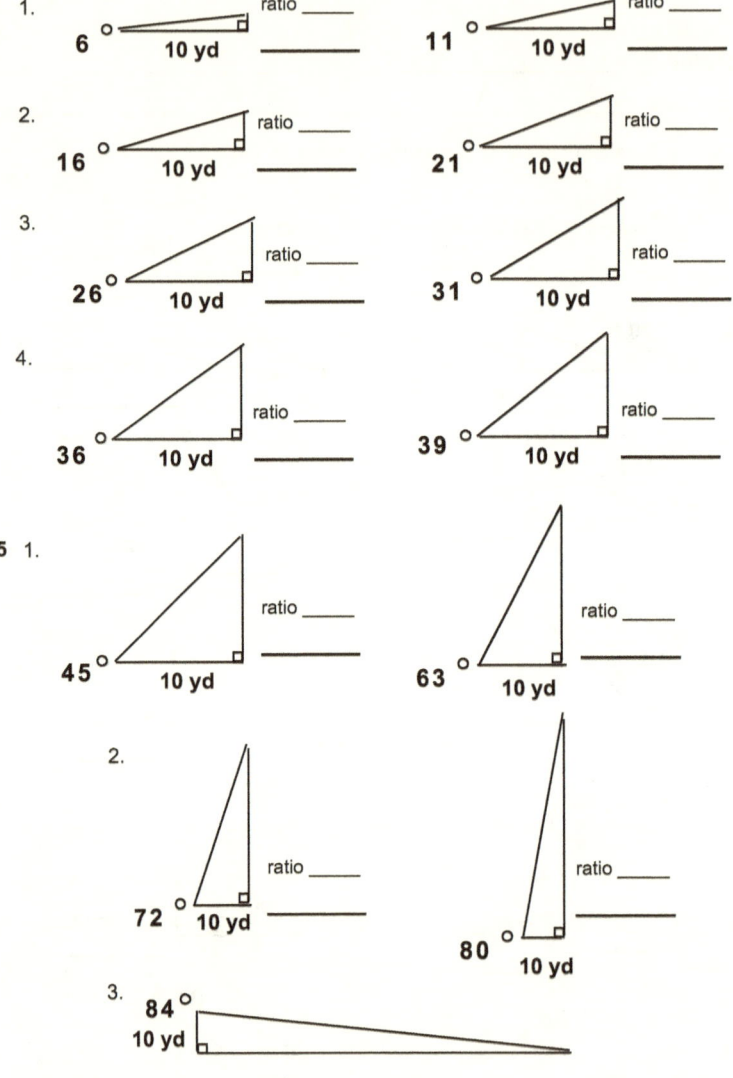

Ch 11 Ls 1 Opposite Ratios for Sine   395

_____ #1 #2 ____/ 6 #3 ____/ 10 R ___/ 5 Total ____/ 21 _____
         Name   Checker

#1 1. What word shows how cosecant, secant, and cotangent are different? _____
_____

2. Which sign is the opposite of sine? _____

3. What is the ratio for cosecant? _____

4. Find the cosecant ratio and a side.

   12° C 10 yd

   _____ yd   10 yd
   12°   ratio _____

5. Find the cosecant ratio and a side.

   24° C 10 yd

   _____ yd   10 yd
   24°   ratio _____

#2 Use Points to Find Ratios

1. This is the sin ratio. What's the cosecant? $\frac{0.5}{1}$ _____

6, 3

What sin and cosine ratio do these graphs make?

(graph with point at (6, 3), axes 0–8 x and 0–6 y)

3 6ths is sin _____ cosine _____

#3 Find the cosecant ratio and side.

Calculator? yes   no

1.
24° 10 yd   ratio _____
18° 10 yd   ratio _____

2.
36° 10 yd   ratio _____
30° 10 yd   ratio _____

3.
6°   10 yd   ratio _____
12° 10 yd   ratio _____

4.
54° 10 yd   ratio _____
45° 10 yd   ratio _____

5.
90° 10 yd   ratio _____
63° 10 yd   ratio _____

Review 1. Which sign is the opposite of sine? _____

Calculator? yes   no

2. What is the ratio for secant? _____

3. What is the ratio for cosecant? _____

4. This is the sin ratio. What's the cosecant? $\frac{0.5}{1}$ _____

5. **What sin and cosine ratio do these graphs make?**

1 5th _____ is sin _____ cosine _____

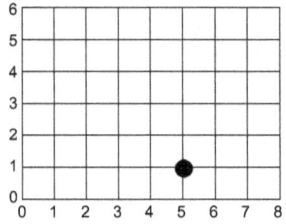

Ch 11 Ls 2 Opposite Ratio for Cosine   397

_____ #1 #2 ____/ 8 #3 ____/10 R ___/ 5 Total ____/ 23 _____
            Name   Checker

#1   1. Which sign is the opposite of cosine? _____
     2. What is the ratio for secant? _____
     3. Here's the cosine ratio. What's the secant? $\frac{0.5}{1}$ _____

     4. Find the secant ratio and a side.

              C
         12° 10 yd

         _____ yd   10 yd
         12°   ratio _____

     5. Find the secant ratio and a side.

              C
         24° 10 yd

         _____ yd   10 yd
         24°   ratio _____

#2   1. What's the sin and cosine $\frac{\pi}{15}$ Radians
        ratio for 1 15th pi radians?

        Angle ____ Sin ratio is _____ Cosine ratio is _____

     2. What's the sin and cosine $\frac{\pi}{6}$ Radians
        ratio for 1 6th pi radians?

        Angle ____ Sin ratio is _____ Cosine ratio is _____

     3. What's the sin and cosine $\frac{\pi}{2}$ Radians
        ratio for 1 half pi radians?

        Angle ____ Sin ratio is _____ Cosine ratio is _____

#3 Find the secant ratio and side.   Calculator? yes no

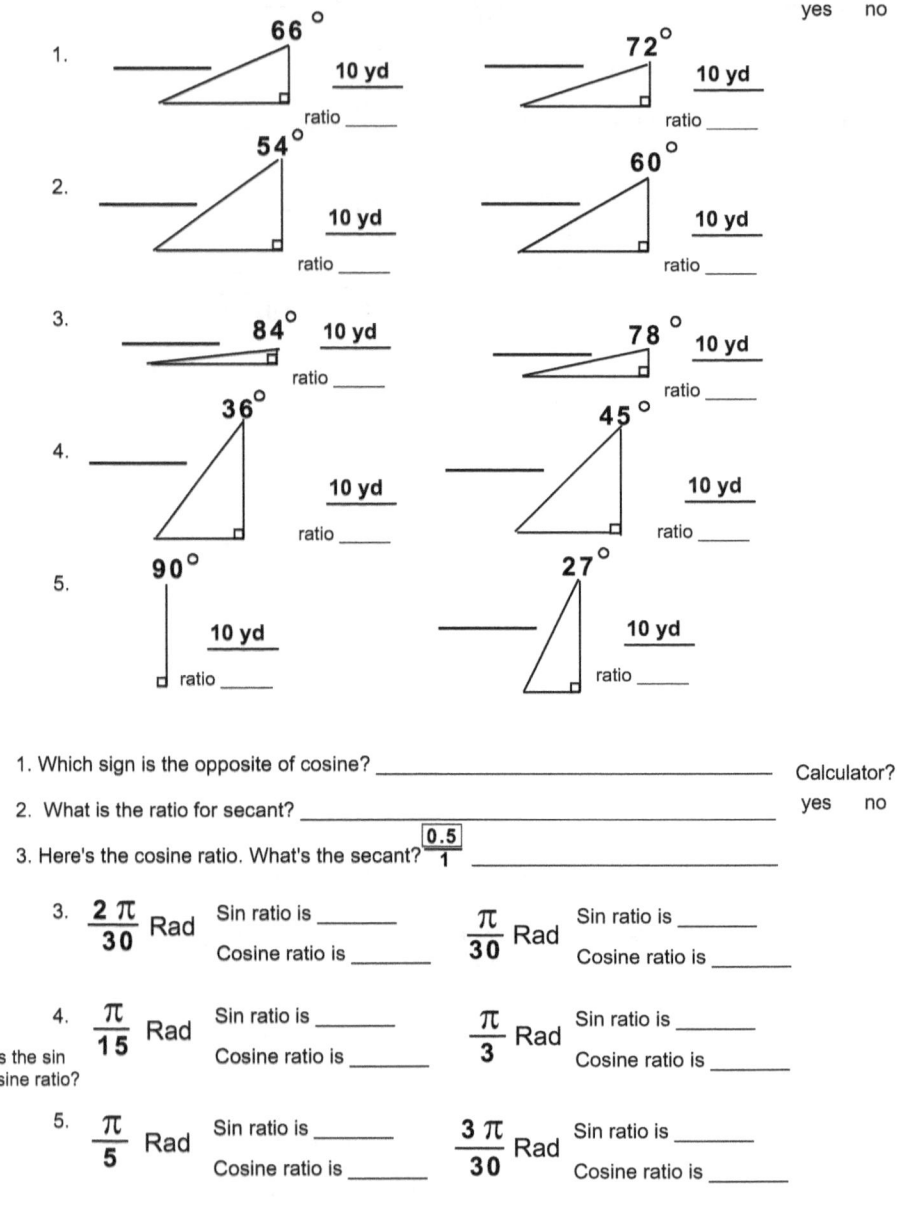

Review 1. Which sign is the opposite of cosine? _____   Calculator? yes no
2. What is the ratio for secant? _____
3. Here's the cosine ratio. What's the secant? 0.5/1 _____

3. $\frac{2\pi}{30}$ Rad   Sin ratio is _____ $\frac{\pi}{30}$ Rad   Sin ratio is _____
                  Cosine ratio is _____   Cosine ratio is _____

What's the sin and cosine ratio?
4. $\frac{\pi}{15}$ Rad   Sin ratio is _____ $\frac{\pi}{3}$ Rad   Sin ratio is _____
                  Cosine ratio is _____   Cosine ratio is _____

5. $\frac{\pi}{5}$ Rad   Sin ratio is _____ $\frac{3\pi}{30}$ Rad   Sin ratio is _____
                 Cosine ratio is _____   Cosine ratio is _____

Ch 11 Ls 3 Opposite Ratio for Tangent   399

_____ #1 #2 ____/10 #3 ____/10 R ___/ 9 Total ____/29 _____
            Name   Checker

#1 1. What is the ratio for tangent? _____

2. What is the opposite ratio of tangent? _____

3. What's the name for opposite of tangent? _____

4. Find the cotangent ratio and a side.   12° a 10 yd

_____ yd   12°   10 yd   ratio _____

5. Find the cotangent ratio and a side.   24° a 10 yd

_____ yd   24°   10 yd   ratio _____

50 m   10 m   x°

#2 1. How does this triangle make a trig equation? _____

2. How does it make a decimal equation? _____

3. What is the sin equation? _____

4. What cosine ratio does 10 50ths show? $\cos x° = 0.70$

$\cos \underline{}° = 0.70$

5. What tangent ratio does 24 30ths show? $\tan x° = 0.21$

$\tan \underline{}° = 0.21$

#3 Find the cotangent ratio and side.   Calculator? yes no

1. 6°, 10 yd — ratio ____ ____
    11°, 10 yd — ratio ____ ____

2. 16°, 10 yd — ratio ____ ____
    21°, 10 yd — ratio ____ ____

3. 26°, 10 yd — ratio ____ ____
    31°, 10 yd — ratio ____ ____

4. 36°, 10 yd — ratio ____ ____
    39°, 10 yd — ratio ____ ____

5. 42°, 10 yd — ratio ____ ____
    45°, 10 yd — ratio ____ ____

Review 1. What is the ratio for tangent? _____   Calculator?
2. What is the opposite ratio of tangent? _____   yes no
3. What's the name for opposite of tangent? _____

4. $\sin x = \frac{10}{30}$ $\sin x = \frac{10}{20}$ $\sin x = \frac{10}{50}$

Answer these problems. _____   _____   _____

5. $\tan x = \frac{10}{15}$ $\tan x = \frac{20}{35}$ $\tan x = \frac{10}{20}$

_____   _____   _____

Review Problems **401**

_____ #1 ____/ 5 #2 ____/ 5   Total ____/ 10
Name

Calculator?
yes   no

#1 1. How tall is a water tower? From a point 60 m from the base of the water tower, you find that you must look up at an angle of 45° to see the top of the tower. How tall is the tower?

Make an equation first.

2. A scientist estimates the heights of features on the moon by measuring the length their shadows. A photograph shows you that a shadow of an out cropping is 300 meters long. When the photograph was taken, the sun's angle to the horizontal surface was 40°. How tall is the outcropping?

3. You lean a ladder 8 meters long against the wall. It makes an angle of 45° with the level ground. How high up is the top of the ladder?

4. How long was the rope for the flagpole? You observe the pole casts a shadow 15 meters long on the ground. The angle between the suns rays and the ground is 38°. How tall is the pole?

5. Your ball is trapped on a tree branch 10 m above the ground. Your ladder is only 8 m long. If you place the ladder's tip on the branch, what angle will the ladder make with the ground?

#2

1. A submarine at the surface of the ocean makes a dive at an angle of 24°. It goes for 600 meters along its downward path. How deep will it be?

_____
_____
_____
_____
_____

Calculator?
yes   no

2. An observer 1 kilometer from the launch pad watches a rocket take off. After 40 seconds the angle of elevation is 24°. How high is the missile at this point?

_____
_____
_____
_____
_____

3. Suppose that the distance from the edge of the creek bed to the edge of the water is 50 meters. The land slopes downward at 32° to the horizontal. How far is the surface of the creek below the level of the surrounding land?

_____
_____
_____
_____

4. You are on a salvage ship. Your sonar system has located a sunken ship at an elevation of depression of 500 meters from your ship, with an angle to the horizontal of 24°. How far must you sail to be directly above the ship?

_____
_____
_____
_____

5. A 600 meter tall tower casts a shadow 800 meters long on the ground. What is the angle of elevation of the sun at that time of day?

_____
_____
_____
_____

Review Problems 403

_____ #1 #2 #3 ___/ 15 #3 #4 ___/ 16 Total ___/ 31
           Name

1. Secant _____
2. Cosecant _____
3. Cotangent _____

#2 Find the cosecant ratio and side.   Calculator?
                                                                yes   no

1.   _____   10 yd   _____   10 yd
      24°   ratio ____   18°   ratio ____

2.   _____   10 yd   _____   10 yd
      36°   ratio ____   30°   ratio ____

3.   _____   10 yd   _____   10 yd
      6°   ratio ____   12°   ratio ____

4.   _____   10 yd   _____   10 yd
      54°   ratio ____   45°   ratio ____

#3 Find the secant ratio and side.   Calculator?
                                                                yes   no

1.   66°
     _____   10 yd   72°
                      ratio ____   _____   10 yd
            54°   ratio ____
2.   60°
     _____   10 yd   _____   10 yd
                      ratio ____   ratio ____

#3 Part 3 Continued

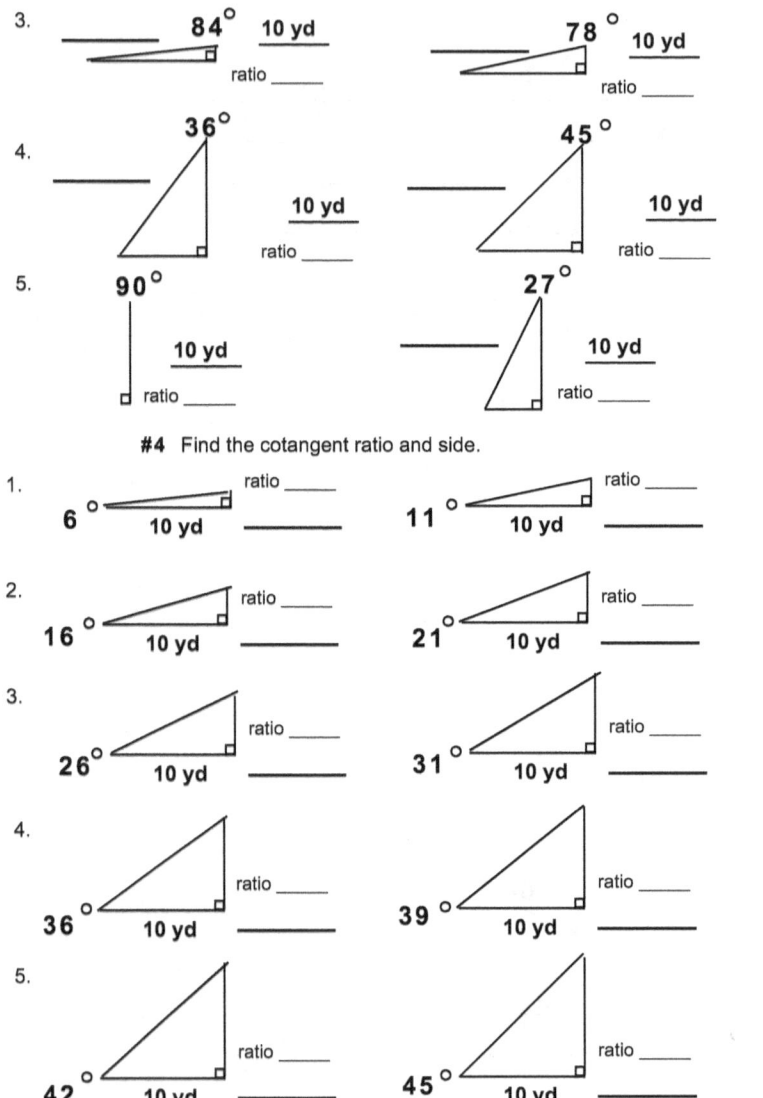

Ch 12 Ls 1 Area of a Triangle   405

_____ #1 #2 ____ / 5 #3 ____ / 3 R ___ / 3 Total ____ / 11 _____
Name   Checker

4 yd
24°
4 yd

#1 1. What formula does sine use to find area? _____
2. What's sin of 24 degrees? _____
3. When do you use 1 half B x C x sin A.? _____

#2 1. What equation solves this?

6 yd
24°
6 yd

What numbers go where?   Area = $\frac{1}{2}$ bc sin A

_____

_____

_____

_____

2. What equation solves this?

8 yd
30°
7 yd

What numbers go where?   Area = $\frac{1}{2}$ bc sin A

_____

_____

_____

_____

#3 Find the areas of these triangles.   Calculator?
                                                           yes   no

1. 14 yd
    30°
       14 yd $\frac{1}{2}$ bc sin A

    _____

    _____

    _____

    _____

2.   10 yd
    24°
       10 yd $\frac{1}{2}$ bc sin A

    _____

    _____

    _____

    _____

3.   5 yd
    18°
       6 yd $\frac{1}{2}$ bc sin A

    _____

    _____

    _____

    _____

       4 yd
    24°
       4 yd

Review 1. What formula does sine use to find area? _____   Calculator?
                                                                                       yes   no
        2. What's sin of 24 degrees? _____

        3. When do you use 1 half B x C x sin A.? _____

Ch 12 Ls 2 What the Law of Sines finds. 407

_____ #1 #2 ____ / 7 #3 ____ / 4 R ____ / 5 Total ____ / 16 _____
            Name   Checker

#1 1. What does the Law of Sines do? _____

2. What does Law of Sines need to work? _____

$12°$ —— 20 —— b
         —— $64°$

3. What finds the angles left? _____

4. What equation solves the ratio? _____

5. What equation solves a proportion? _____

#2 1. Use Sine Law. What proportion finds **a**?

   $30°$ — a — 10 inches — $24°$

Solve with cross multiply.   sin ____ = sin ____

Get the variable by itself.   _____ = _____

Find the sin ratios.   _____

Solve for a.   _____

2. Use Sine Law. What proportion finds **a**?

   $36°$ — a — 15 inches — $24°$

Solve with cross multiply.   sin ____ = sin ____

Get the variable by itself.   _____ = _____

Find the sin ratios.   _____

Solve for a.   _____

#3 Use the law of sines.   Calculator? yes   no

1.

   5 a
36°△24°

sin____ = sin____

_____
_____
_____
_____

   6 a
12°△53°

sin____ = sin____

_____
_____
_____
_____

2.

   30 a
24°△12°

sin____ = sin____

_____
_____
_____
_____

   15 a
36°△30°

sin____ = sin____

_____
_____
_____
_____

Review 1. What does the Law of Sines do? _____   Calculator?
2. What does Law of Sines need to work? _____   yes   no

   20 b
12°△64°

3. What finds the angles left? _____
4. What equation solves the ratio? _____
5. What equation solves a proportion? _____

(Triangles not drawn to scale).

Ch 12 Ls 3 Triangle Signs   409

_____ #1 #2 ____ / 28 #3 ____ / 4 R ___ / 4 Total ____ / 36 _____
           Name   Checker

#1 1. What does "other angles" stand for? _____
      2. Which quadrant has all 3 signs positive? _____
      3. What is a standard triangle? _____
      4. What letters show which quadrants have positive signs? _____

#2 1. Find sin. Is it positive or negative?

Triangle: 24°, sides 1, 0.9, 0.4   rounded

Positive   Negative

                                                                    Add 90 degrees. _____ Is it positive or negative?

Positive   Negative   90 + _____ = _____   Add 180 degrees. Is it positive or negative?

Positive   Negative   180 + _____ = _____   Add 270 degrees. Is it positive or negative?

Positive   Negative   270 + _____ = _____

2. Find cosine. Is it positive or negative?

Triangle: 24°, sides 1, 0.9, 0.4

Positive   Negative

                                                                     Add 90 degrees. _____ Is it positive or negative?

Positive   Negative   90 + _____ = _____   Add 180 degrees. Is it positive or negative?

Positive   Negative   180 + _____ = _____   Add 270 degrees. Is it positive or negative?

Positive   Negative   270 + _____ = _____

#3 Is each function positive, negative, zero, or undefined.   Calculator?
                                                                yes   no

1. **sin 120 d**   **cos 405 d**   **tan 180 d**
    Pos Neg 0 Und   Pos Neg 0 Und   Pos Neg 0 Und

2. **sin (- 120 d)**   **cos 270 d**   **tan 360 d**
    Pos Neg 0 Und   Pos Neg 0 Und   Pos Neg 0 Und

3. **sin (- 180 d)**   **cos 405 d**   **tan (- 90 d)**
    Pos Neg 0 Und   Pos Neg 0 Und   Pos Neg 0 Und

4. **sin (- 270 d)**   **cos (- 189 d)**   **tan 320 d**
    Pos Neg 0 Und   Pos Neg 0 Und   Pos Neg 0 Und

#4 Find the corresponding angles.

1. **sin 45 d**   **cos 180 d**   **tan 270 d**

    _____   _____   _____

2. **sin 135 d**   **cos 225 d**   **tan 320 d**

    _____   _____   _____

3. **sin (- 150 d)**   **cos (- 270 d)**   **tan (- 150 d)**

    _____   _____   _____

Review 1. What does "other angles" stand for? _____

2. Which quadrant has all 3 signs positive? _____

3. What is a standard triangle? _____

4. What letters show which quadrants have positive signs? _____

Ch 12 Ls 4 Law of Cosines   411

_____ #1 #2 ____/ 28 #3 ____/ 4 R ___/ 4 Total ____/ 36 _____
           Name   Checker

#1 1. When do you use Law of Cosines? _____
    2. What's the left side of Law of Cosines? _____
    3. What else is multiplied? _____
    4. What does the 1st letter show? _____

#2 1. What equation solves this?

                                        6
                                    26   a
                            A
                                      6

    What numbers go where? $a^2 = b^2 + c^2 - 2bc \cos A$

    _____

    _____

    _____

    _____

2. What equation solves this?

                                        5
                                  26   a
                          A
                                5

    What numbers go where? $a^2 = b^2 + c^2 - 2bc \cos A$

    _____

    _____

    _____

    _____

#3 1. 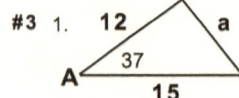 $a^2 = b^2 + c^2 - 2bc \cos A$ Calculator?
yes   no

_____
_____
_____
_____
_____

2. 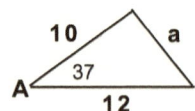 $a^2 = b^2 + c^2 - 2bc \cos A$

_____
_____
_____
_____
_____

3.

$a^2 = b^2 + c^2 - 2bc \cos A$

_____
_____
_____
_____
_____

Review 1. When do you use Law of Cosines? _____
2. What's the left side of Law of Cosines? _____
3. What else is multiplied? _____
4. What does the 1st letter show? _____

Ch 12 Ls 5   Centripetal Acceleration   413

_____   #1 #2 ____/ 8   #3 ____/ 6   R ___/ 6   Total ____/ 20 _____
         Name   Checker

#1   1. What does centripetal acceleration find? _____
          2. What is the formula for centripetal acceleration? _____
          3. What does A stand for in centripetal acceleration? _____
          4. What kind of answer does it find? _____

#2   1. What formula is it?   **Centripetal Acc 3 m/s/s   Radius 2 meters**

          Which numbers go where?   _____

          Solve the equation.   _____

                                   _____

          2. Make an equation.   **Centripetal Acc 20 cm/s/s   Radius 30 cm**

          Which numbers go where?   _____

          Solve the equation.   _____

                                   _____

          3. Make an equation.   **Centripetal Acc 80 cm/s/s   Radius 40 cm**

          Which numbers go where?   _____

          Solve the equation.   _____

                                   _____

          4. What is happening? $\sqrt{(20)5} = v$
                                    m/sec² m

                                   _____

#3 1. $\sqrt{(10)5} = v$
m/sec² m

$\sqrt{(40)7} = v$
m/sec² m

_____

_____

_____

_____

2. $\sqrt{(20)r} = 12$
m/sec² m

$\sqrt{(24)r} = 14$
m/sec² m

_____

_____

_____

_____

_____

_____

3. $\sqrt{(a)10} = 30$
m/sec² m

$\sqrt{(a)20} = 35$
m/sec² m

_____

_____

_____

_____

_____

_____

Review 1. What does centripetal acceleration find? _____
2. What is the formula for centripetal acceleration? _____
3. What does A stand for in centripetal acceleration? _____
4. What kind of answer does it find? _____
5. $\sqrt{(90)5} = v$
m/sec² m

$\sqrt{(120)12} = v$
m/sec² m

_____

_____

_____

_____

Review Problems 415

_____ #1 ____ / 5 #2 ____ / 5 Total ____ / 10
              Name

                                                                    Calculator?
                                                                    yes   no

#1 1. A boat is sailing east at a speed of 18 kph. A   _____
       lighthouse is 70 degrees south, and 20 min
       later it's 50 degrees south. Find the distance   _____
       from the boat to the lighthouse.
                                                            _____
Make an equation
before solving.   _____

    2. Two planes leave an airport at the same time.   _____
       One plane is flying east at 500 kph. The other
       plane is flying north for 550 kph. How far apart   _____
       are the planes apart after flying for 2 hours?
                                                            _____

                                                            _____

    3. A poll tilts towards the sun at a 12 degree angle   _____
       from the vertical at it casts a 8 m shadow. The
       angle of elevation from the shadow to the top   _____
       of the pole is 45 degrees. How tall is the poll?
                                                            _____

                                                            _____

    4. To determine the length of a lake a surveyor   _____
       starts at Point A and walks to a position
       that is 500 meters North of the staring point to   _____
       Point B. The surveyor then walks to Point C,
       which is 300 meters South of the 2nd Point.   _____
       The angle formed by Segment AB and BC is
       120 degrees. How long is the pond?   _____

    5. A post is supported by two wires (one on each   _____
       side going in opposite directions) creating an
       angle of 100° between the wires. The ends of   _____
       the wires are 14m apart on the ground with one
       wire forming an angle of 45° with the ground.   _____
       Find the lengths of the wires.
                                                            _____

#2 Story Problems

Calculator? yes   no

1. Ojas and Amav both start at point A. They each walk in a straight line at an angle of 110° to each other. After 50 minutes Ojas has walked 4 km and Amav has walked 3 km. How far apart are they?

2. Two ships are sailing from London. #1 is sailing due east and #2 is sailing 50° south of east. After an hour, #1 has travelled 130 km and the #2 has travelled 100 km. How far apart are the two ships?

3. Points A and B are on opposite sides of a valley. Point C is 300 meters from A. Angle B measures 90° and angle C measures 60°. What is the distance between A and B?

4. A fire is spotted from two lookout stations that are 20 km apart. The bearing from the first lookout station to the fire is N 60 degrees E. The bearing from the second lookout station to the fire is N 30 degrees W. Find the distance from each lookout station to the fire.

5. Two jets leave an airport at the same time. One plane is flying 900 kph at a bearing N 30 degrees E, and the other jet is flying at 900 kph at a bearing of N 70 degrees W. How far apart are the planes after flying for 3 hours?

Review Problems 417

_____ #1 #2 #3 ____ / 7 #4 #5 ____ / 4 Total ____ / 11
                Name

1. Area of a Triangle _____

2. Law of Sines _____

3. Law of Cosines _____

#2 Find the areas of these triangles.   Calculator?
                                                       yes   no

1.   14 yd
   30°
         14 yd $\frac{1}{2}$ bc sin A

   _____
   _____
   _____
   _____

2.   10 yd
   24°
         10 yd $\frac{1}{2}$ bc sin A

   _____
   _____
   _____
   _____

#3 Use the law of sines.   Calculator?
1.   yes   no
     5   a   6   a
   36°   24°   30°   18°

   sin ____ = sin ____   sin ____ = sin ____

   _____   _____
   _____   _____
   _____   _____
   _____   _____

#4 Find the corresponding angles.   Calculator? yes no

1. Name the next 3 angles.
   (Add 90, 180, and 270)

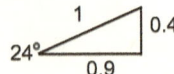

| Find sin. Positive or negative? | 24° | 114° | 204° | 294° |

_____ _____ _____ _____

2. Name the next 3 angles.
   (Add 90, 180, and 270)

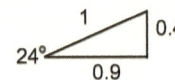

| Find sin. Positive or negative? | 24° | 114° | 204° | 294° |

_____ _____ _____ _____

#5 Use the angles to find the corresponding angles.   Calculator? yes no

1. 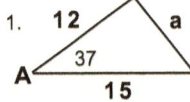 $a^2 = b^2 + c^2 - 2bc \cos A$

_____
_____
_____
_____

2. 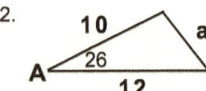 $a^2 = b^2 + c^2 - 2bc \cos A$

_____
_____
_____
_____

Ch 13 Ls 1 Circular Functions 419

_____ #1 #2 ____/ 8 #3 ____/ 3 R ___/ 5 Total ____/16 _____
Name   Checker

#1 1. How are circular functions like a ferris wheel? _____
    2. What does the graph use for X's? _____
    3. What does sine graph look like? _____
    4. How is the cosine graph different? _____
    5. How does it find an angle? _____
    6. Find these points.

    0.5 pi   1.0 pi   2.0 pi   3.5 pi

    ____   ____   ____   ____

#2 1. Find the degrees for this angle thru 360.   **390 degrees**

                                                      ____

    2. Find the degrees for this angle thru 360.   **420 degrees**

                                                      ____

    3. Find the degrees for this angle thru 360.   **390 degrees**

                                                      ____

    4. Find the degrees for this angle thru 360.   **420 degrees**

                                                      ____

    5. Find the degrees for this angle thru 360.   **420 degrees**

                                                      ____

                                         I apologize for the graph.

420

#3 Find these points from the graphs.   Calculator? yes   no

#2 1. Find these points.

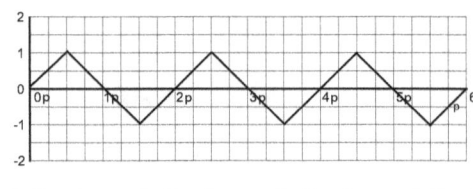

1.5 pi   1.5 pi   1.5 pi   1.5 pi

_____   _____   _____   _____

Review 1. How are circular functions like a ferris wheel? _____   Calculator? yes   no

2. What does the graph use for X's? _____

3. What does sine graph look like? _____

4. How is the cosine graph different? _____

5. How does it find an angle? _____

6. Find the degrees for this angle thru 360.   **sin 390 degrees**

_____ degrees

7. **sin 420 degrees**   **sin 450 degrees**   **sin 480 degrees**

_____ degrees   _____ degrees   _____ degrees

8. **sin 510 degrees**   **sin 540 degrees**   **sin 570 degrees**

_____ degrees   _____ degrees   _____ degrees

9. **sin 600 degrees**   **sin 630 degrees**   **sin 660 degrees**

_____ degrees   _____ degrees   _____ degrees

10. **sin 690 degrees**   **sin 720 degrees**   **sin 750 degrees**

_____ degrees   _____ degrees   _____ degrees

Ch 13 Ls 2 Graph Tangential Functions   421

_____ #1 #2 ____ / 6 #3 ____ / 3 R ___ / 3 Total ____ / 12 _____
Name   Checker

#1 1. How is tangent graph different? _____

2. How is the tangents multigraph different? _____

3. What do you need to remember about negative numbers? _____

_____

4. What are these degrees?   **1.5 pi**   **1.5 pi**   **1.5 pi**   **1.5 pi**

_____   _____   _____   _____

Use thise graph for
these questions.

#2 1. Find the radian for this angle.   **405 degrees**

_____

2. Find the radian for this angle.   **450 degrees**

_____

3. Find the radian for this angle.   **540 degrees**

_____

4. Find the radian for this angle.   **630 degrees**

_____

5. Find the radian for this angle.   **675 degrees**

_____

#3 1. Find these points.

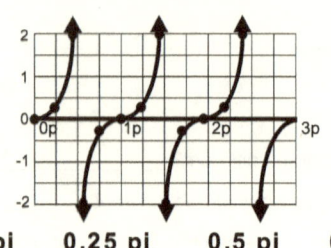

0 pi   0.25 pi   0.5 pi   0.75 pi

_____   _____   _____   _____

Calculator? yes no

2. Find these points.

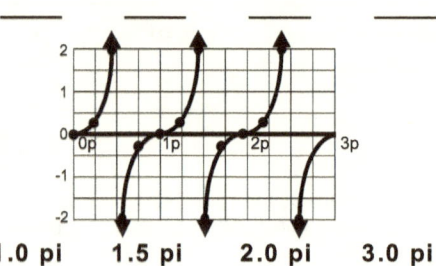

1.0 pi   1.5 pi   2.0 pi   3.0 pi

_____   _____   _____   _____

Review 1. $y = \sin 2q$ How did this graph change? _____

2. $y = 2 \sin q$ How did this graph change? _____

3. How do you tell the difference? _____

Calculator? yes no

6. Find the degrees for this angle thru 360.   **sin 390 degrees**

_____ degrees

7. Find the degrees. **sin 420 degrees**   **sin 420 degrees**

_____ degrees   _____ degrees

8. Find the degrees. **sin 390 degrees**   **sin 390 degrees**

_____ degrees   _____ degrees

9. Find the degrees. **sin 420 degrees**   **sin 420 degrees**

_____ degrees   _____ degrees

Ch 13 Ls 3   Change the Graph

_____ #1 #2 ____ / 6 #3 ____ / 3 R ____ / 3 Total ____ / 12 _____
         Name   Checker

#1 1. If the graph goes twice as fast, what did it change? _____

2. How does a sentence show this? _____

3. The graph is twice as tall. What changed this time? _____

4. How does a sentence show this ? _____

5. **y = sin 2q** How did this graph change? _____

6. **y = 2 sin q** How did this graph change? _____

7. How do you tell the difference? _____

8. Find these points.   **0 pi**   **0.5 pi**   **1.0 pi**   **1.5 pi**

          ____   ____   ____   ____

#2 1. Find the radian for this angle. **390 degrees**

                          _____ radians

2. Find the radian for this angle. **420 degrees**

                          _____ radians

3. Find the radian for this angle. **480 degrees**

                          _____ radians

4. Find the radian for this angle. **510 degrees**

                          _____ radians

5. Find the radian for this angle. **570 degrees**

                          _____ radians

#3

1.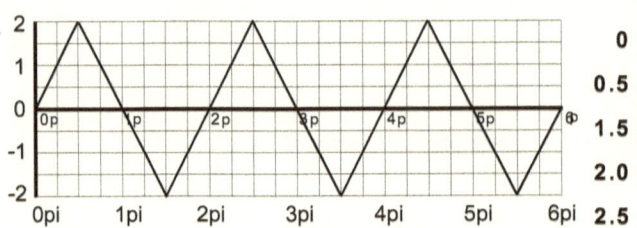

0 pi = _____
0.5 pi = _____
1.5 pi = _____
2.0 pi = _____
2.5 pi = _____

Estimate these points.

What equation is it? _____

2.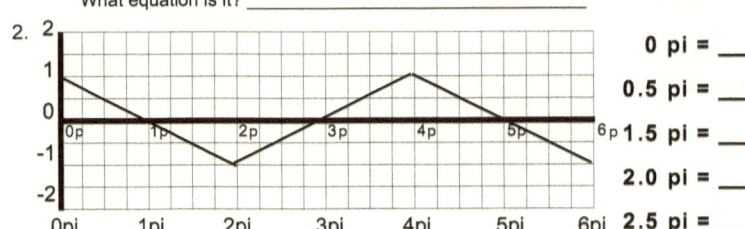

0 pi = _____
0.5 pi = _____
1.5 pi = _____
2.0 pi = _____
2.5 pi = _____

What equation is it? _____

3.

0 pi = _____
0.5 pi = _____
1.5 pi = _____
2.0 pi = _____
2.5 pi = _____

What equation is it? _____

Review

1. If the graph goes twice as fast, what did it change? _____

Calculator?
yes   no

2. How does a sentence show this? _____

3. The graph is twice as tall. What changed this time? _____

4. How does a sentence show this? _____

5. $y = \sin 2q$ How did this graph change? _____

6. $y = 2 \sin q$ How did this graph change? _____

7. How do you tell the difference? _____

Ch 13 Ls 3 Change the Graph Continued 425

_____ #1 #2 ____/ 7 #3 ____/ 3 R ___/ 4 Total ____/ 14 _____
            Name   Checker

#1 1. If the graph goes twice as fast, what changed? _____
    2. How does a sentence show this? _____
    3. The graph is twice as tall. What changed this time? _____
    4. Where does sin equation change the amplitude? _____
    5. How do you remember period and amplitude signs? _____

#2 1. Find the degrees.   **390 sin degrees**   **420 sin degrees**

                         _____ degrees is _____   _____ degrees is _____

    2. Find the degrees.   **450 sin degrees**   **480 sin degrees**

                         _____ degrees is _____   _____ degrees is _____

    3. Find the degrees.   **510 sin degrees**   **540 sin degrees**

                         _____ degrees is _____   _____ degrees is _____

    4. Find the degrees.   **390 cos degrees**   **420 cos degrees**

                         _____ degrees is _____   _____ degrees is _____

    5. Find the degrees.   **450 cos degrees**   **480 cos degrees**

                         _____ degrees is _____   _____ degrees is _____

    6. Find the degrees.   **510 cos degrees**   **540 cos degrees**

                         _____ degrees is _____   _____ degrees is _____

    7. Find the degrees.   **390 tan degrees**   **420 tan degrees**

                         _____ degrees is _____   _____ degrees is _____

#3   1.   **sin 570 d**   **sin 600 d**   **sin 630 d**   Calculator?
                                                                          yes   no
        ____ degrees is ____   ____ degrees is ____   ____ degrees is ____

    2.   **sin 660 d**   **sin 690 d**   **sin 720 d**
        ____ degrees is ____   ____ degrees is ____   ____ degrees is ____

    3.   **cos 570 d**   **cos 600 d**   **cos 630 d**
        ____ degrees is ____   ____ degrees is ____   ____ degrees is ____

    4.   **cos 660 d**   **cos 690 d**   **cos 720 d**
        ____ degrees is ____   ____ degrees is ____   ____ degrees is ____

    5.   **tan 450 d**   **tan 480 d**   **tan 510 d**
        ____ degrees is ____   ____ degrees is ____   ____ degrees is ____

    6.   **tan 540 d**   **tan 570 d**   **tan 600 d**
        ____ degrees is ____   ____ degrees is ____   ____ degrees is ____

Review 1. If the graph goes twice as fast, what changed? _____   Calculator?
                                                                                    yes   no
        2. How does a sentence show this? _____
        3. The graph is twice as tall. What changed this time? _____
        4. Where does sin equation change the amplitude? _____
        5. How do you remember period and amplitude signs? _____
        _____

        6. **A pilot is flying from A to B, a distance of 400**   _____
           kilometers. In order to avoid thunderstorms
           they alter their course by 20 degrees and fly   _____
           80 kilometers. How far are they from airport B?
                                                                    _____
           80 km
              20 deg   _____
           A   400 kilometers   B
                                                                    _____

Review Problems 427

_____ #1 #2 ____/10 #3 #4 ____/15 R ___/14 Total ____/39
        Name

#1 1. **Circular Functions** _____
       2. **Period in Trigonometry** _____
       3. **Amplitude** _____

#2 1. Find the degrees.   **sin 420 degrees**   **sin 420 degrees**

                               _____ degrees   _____ degrees

       2. Find the degrees.   **sin 390 degrees**   **sin 390 degrees**

                               _____ degrees   _____ degrees

       3. Find the degrees.   **sin 420 degrees**   **sin 420 degrees**

                               _____ degrees   _____ degrees

       4. Find the degrees.   **sin 420 degrees**   **sin 420 degrees**

                               _____ degrees   _____ degrees

#3 Do these word problems on circular functions.

1. **The motion of a weight on a spring is pulled down 10 centimeters from it's equilibrium point and then released. It bounces up to a point and returns back to the release point in 2 seconds. Graph 3 turns of it.**

#4 Estimate these points.

1.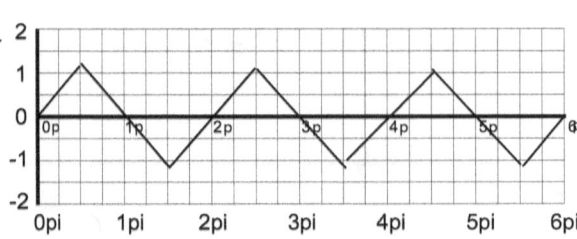

0 pi = _____ Calculator? yes no
0.5 pi = _____
1.5 pi = _____
2.0 pi = _____
2.5 pi = _____

What equation is it? _____

2.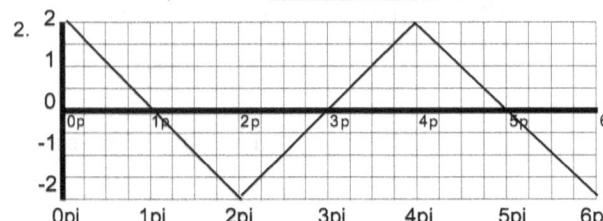

0 pi = _____
0.5 pi = _____
1.5 pi = _____
2.0 pi = _____
2.5 pi = _____

What equation is it? _____

3.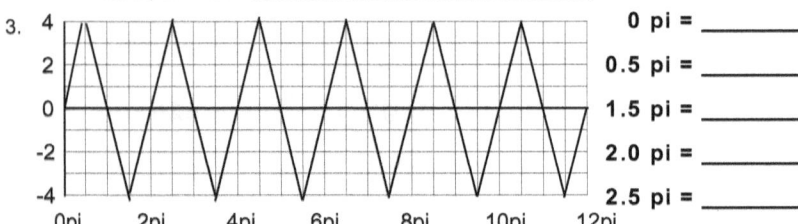

0 pi = _____
0.5 pi = _____
1.5 pi = _____
2.0 pi = _____
2.5 pi = _____

What equation is it? _____

4.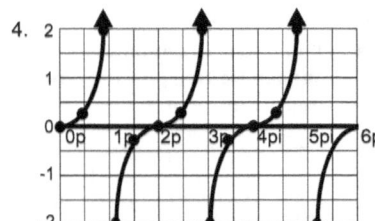

0 pi = _____
0.5 pi = _____
1.5 pi = _____
2.0 pi = _____
2.5 pi = _____

What equation is it? _____

www.ingramcontent.com/pod-product-compliance
Lightning Source LLC
Chambersburg PA
CBHW020720180526
45163CB00001B/48